DISTANT VOICES NEAR

DISTANT VOICES NEAR

HISTORICAL GLOBALIZATION
AND INDIAN RADIO IN
TRINIDAD AND TOBAGO

Shaheed Nick Mohammed

THE UNIVERSITY OF THE WEST INDIES PRESS
Jamaica · Barbados · Trinidad and Tobago

The University of the West Indies Press
7A Gibraltar Hall Road, Mona
Kingston 7, Jamaica
www.uwipress.com

A catalogue record of this book is available from the
National Library of Jamaica.

ISBN: 978-976-640-639-4 (print)
978-976-640-640-0 (Kindle)
978-976-640-641-7 (ePub)

Cover image: Fort in Jaipur, India, over Port of Spain, Trinidad skyline.
Photos and composite by Shaheed Nick Mohammed
Cover and book design by Robert Harris
Set in Minion Pro 10.5/14.5 x 24
Printed in the United States of America

For my dear departed father, Shaffie Mohammed,
whose curiosity and love of knowledge live on in all of us.

Contents

List of Abbreviations *ix*

Introduction *1*

1   Trinidad and Tobago: Historical-Global Developments *15*

2   Culture and Globalization *39*

3   Early Radio in the Caribbean *53*

4   Media and Indian Content in Trinidad and Tobago *76*

5   103 FM and Indian Radio Stations in Trinidad and Tobago *93*

6   Media, Religions and Radio Jaagriti *106*

7   The "Corporate" Mainstream *121*

8   Hybrid Visions: WIN 101, Heritage and Shakti *128*

9   The Global Dimensions *139*

Conclusion: Contextualizing Indian Radio *157*

References *177*

Index *191*

Abbreviations

AM	amplitude modulation
BBC	British Broadcasting Corporation
FM	frequency modulation
HCU	Hindu Credit Union
PDP	People's Democratic Party
PNM	People's National Movement
SDMS	Sanatan Dharma Maha Sabha
TBC	Trinidad Broadcasting Company
TTT	Trinidad and Tobago Television

Introduction

"My radio dial stuck 'pon Irie FM – and you know what? Me nah bother fix it!"

These were the words of one of the original promos from Jamaica's Irie FM radio station, the first ever twenty-four-hour all-reggae station to be launched. In the island that produced and nurtured reggae music, the idea of a specialist reggae radio station was still a novelty in 1990, when Irie was launched. Indeed, the relationship between radio and reggae had not always been the symbiosis that one might imagine. There were cultural struggles over what was considered legitimate fare for broadcast, and local media were often accused of being biased against local music. Following the loosening of government restrictions on private radio broadcasting licences, Irie quickly established itself as a major force in Jamaican radio despite widespread concerns about whether the market would support it. While Irie was a commercial station, the historical and cultural forces that surrounded it elevated the station to a position of pride and saw it valued as an expression of Jamaican identity.

This book, however, is not about Irie FM. Rather, this work deals with an accumulation of globalized cultural tensions that have come to be embodied not too far away from Jamaica, in the small two-island nation of Trinidad and Tobago. In particular, the present work considers how these tensions, developed through globalized exchanges of people and ideas over many years, have come to be embodied in one particular unique commercial mass-media phenomenon – Trinidad's Indian radio.

The establishment in the early 1990s of a single station for broadcasting music and other programming aimed at the island's population of histori-cal descendants of Indian immigrants was the catalyst for a slew of imitators and the development of an entire sector of the broadcast industry catering to a particular ethno-cultural audience. This sector would diversify over the first two decades, with the evolution of stations that, on the one hand, would compartmentalize into particular niches such as super-specialized "Indian" religious programming and, on the other, some that would generalize into a broader offering of a mix of cultural styles and hybrid music.

After a history of over twenty years, the Indian radio sector or Indian-format radio in Trinidad and Tobago has carved a clear swath of territory in the local market. Alongside a wider assortment of radio offerings ranging from news and talk to evangelical, as well as urban, adult contemporary and local/regional for-mats, Indian radio provides both a backdrop to national life for its target com-munity and a focal point of cultural attention. The broader electronic media market features seven national television stations, ten cable-only television broadcasters and thirty-eight radio stations, in addition to multiple options for cable television reception and widespread internet and mobile penetration.

In the face of such external competition for audiences, and ongoing internal struggles to secure listeners, Trinidad's Indian-format radio stations have been forced to innovate both creatively and technically. These challenges for securing listenership have seen, for example, the promotion of events including concerts featuring foreign artistes and competitions showcasing local talent. To increase listenership, these stations were also quick to innovate into Internet streaming and to funnel song requests to social media. Thus their audiences engage both locally and abroad with their traditional broadcast signals, Internet streams and social media platforms.

While the Indian-format stations continue to face challenges in being accepted into the cultural mainstream, they have continued to be popular with their listeners and advertisers. These stations are played, often loudly, at family gatherings both in Trinidad and in Indo-Trinidadian diaspora communities in the United States, Canada and elsewhere. The stations also serve as a nexus for family messaging through song dedications, both among family members within Trinidad and Tobago and, increasingly, as a kind of trans-national fam-ily musical-dedication and messaging service through social media.

Competition and attrition have seen several movements of stations in and

out of the sector and numerous movements of management and personalities among the stations, but for nearing a quarter of a century, these stations,
whether independent or part of a media conglomerate, have created a significant media phenomenon. That being said, beyond its scale and scope, this
particular industry sector is not unique. There are many examples of so-called
ethnic media in many parts of the world – often tied to minority or plural communities. Singapore, for example, features several commercial radio stations
programmed in the languages of its component linguistic groups.

Indian radio in Trinidad, however, does present some interesting features
that distinguish it from other similar media enterprises. The very presence of
Indian music radio in Trinidad and Tobago suggests notions of hybridity in
that a cultural form from far away (with, arguably, only a vestigial connection to the local environment) finds both expression and commercial success.
Indeed, the content of some of Trinidad's Indian radio (no such station exists
in Tobago) programming that includes American, Jamaican and Trinidadian
music as well as mash-ups or mixes with Indian songs further suggests that
they are hybrid cultural artefacts, combining two or more cultural forms. Yet
hybridity as a formal explanatory framework may miss the subtleties of the
emergent, eclectic, and often deliberate nature of some of these cultural interweavings.

As I examine in the following chapters, cultural artefacts – even modern ones such as Trinidad and Tobago's Indian-radio media sector – reflect
complex interrelationships among many forces. Among these are historical
factors, including global forces such as British colonialism, which brought to
the islands not only the Indian indentured labourers but also African slaves
before them, the European plantation owners and several smaller populations
of imported labour. The resulting globalized environment faced challenges
surrounding other forces such as language, culture and religion, all of which
served as sources of difference and, at times, the sites of contestation and negotiated meanings.

Simple generalizations about the roles of colonialism, language, culture
and religion are often difficult to make, since these often worked in multiple
and complex ways, given the various tensions of the indentureship and post-
indentureship Trinidad and Tobago society. In particular, the emerging politics
of self-determination towards the end of the colonial project exposed rifts not
only between the descendants of African slaves and the descendants of Indian

indentured labourers, but also within and among these groups. The cultural relationships between and within these groups became strained as they negotiated their positions relative to one another and relative to the colonial project. Part of this was their negotiation of positions relative to the place in which they found themselves – an outpost of the British Empire – and how all of that would change with the emergent notion of Trinidad and Tobago as a nation.

After Trinidad and Tobago gained independence from Britain, nationalisms in their broadest sense continued to play some role – though these were primarily (though not only) at the level of social discourse and the struggle for ideological independence. Such struggles were embodied and perhaps taken to their most intense forms in movements such as the Black Power movement of the 1970s, identified with the leadership of several African-oriented groups as well as key trade unions. Yet the notion that the Black Power movement excluded Indo-Trinidadians is probably not completely true, since certain key players in the most radical and violent activities of the movement and in some of its leadership and support roles were Indo-Trinidadian. Also, a number of the movement's initiatives involved efforts to unify and mobilize workers in traditionally Indo-Trinidadian enclaves to engage in the struggle.

Cultural activist Ravindranath Maharaj (Ravi Ji), who has been involved in efforts to revive and clarify Indo-Trinidadian Hindu traditions, spoke of his reactions to the Black Power movement in private, informal conversation with the author, but also shared the same sentiments with a local newspaper (Peter Ray Blood, "Black Power: A Much Needed Revolution", *Trinidad and Tobago Guardian*, 22 April 2015, 14):

> When the Black Power hit the country I was very, very scared, hearing about these Africans coming to Central to join sugar cane workers. I subsequently felt some relief when Bhadase (Sagan Maraj) made a statement giving the assurance that the marchers would not actually enter the canefields. . . . Hearing of the march to Caroni I must admit that I had a sense of concern. But, I was curious so I went to Chaguanas to see the march. When I got there my fear dissipated somewhat as, instead of seeing militant agitators, I remember what I saw were many thirsty, tired young people sitting around. At that time I was close to Gerald Bryce, an official then of the Black Panther movement. . . . When I returned in 1983 I reconnected to the Black Power movement. It's ironic because I actually went to India with two dashikis. Because of Bryce I became closer to the Black Power movement. One of the things I realized was the impact of the movement opened the way for transformation because there

were obvious changes in the employment practices of the country. In the banks you
could have easily discerned change (by the ethnicity of the people being employed).

Questions of cultural identity are not, however, only evident in overt political struggles. They are often also at play in some of the most mundane aspects
of everyday life. Several scholars and commentators have noted, for example,
that for many years Indo-Trinidadian cricket fans would display their support
for teams from India and Pakistan instead of their home team, the West Indies.
This particular pattern of support has also been observed in Guyana.

The particular dynamics of Trinidad and Tobago's ethno-cultural relationships have engaged the attention of many scholars over the years. For many
reasons, including the generally peaceful (if contentious) coexistence of the two
main ethnic groups and several smaller groups, various commentators have
attempted to make sense of the cultural and the national in Trinidad. They have
variously experimented with notions of hybridities, cultural mixings, melting
pots and even callaloo as metaphors for Trinidad and Tobago's culture. As we
shall explore here, some analyses have been overtly political and frequently
suffer from both cultural insularity and ethnic rancour. Indeed, many of the
most prominent commentators on Trinidad society include some of the most
ethnically biased and myopic personalities (some of whom are academics and
religious leaders).

In pursuit of this divisive agenda, some commentators (academics among
them) have tended to overstate questions of culture, heritage and national identity in terms of the divisive nationalisms that have been evident in many other
countries. This tradition of couching cultural expression as nationalistic or
separatist in some way harks back to Eric Williams and his political machin-
ations to discredit his Indo-Trinidadian opponents. These opponents played
into his hand in no small way by using names reminiscent of Indian nationalist
groups and also by actions such as instituting the singing of the Indian national
anthem in their schools following Indian independence (but prior to Trinidad
and Tobago's independence).

Post-independence, there is evidence of competing tensions between cultural heritage on the one hand and national identity on the other, at least for
those whose cultural heritage is defined as outside the national mainstream.
This raises important questions about the very definitions of "givens" such as
culture and whether concepts such as identity and heritage are as fixed or as

pure as we are taught to believe. By interrogating the very roots of these various elements of social and personal identity, and exposing their malleable nature, we may also be able to challenge the idea that these are essential, fixed or even relevant in the modern world.

At the same time, while it is necessary to lift the anchor of essentiality from historic, globalized identity claims to expose them as caprices of history and traditions of ignorance (often imagined into being), it is also important to examine the extent to which modern identities can be constructed and imagined out of modern global forces, including mass media with international reach. In the case of Trinidad and Tobago, this presents a particularly difficult set of entanglements – with often misremembered and generationally morphed traditions mixed with influences from old Indian films, as well as more recent involvements with Indian soap operas and commercial experiences of Indian traders. Into this complex mix of cultural involvements we must add the Indian radio station, whose connections with India are sometimes tenuous. The music played on these stations may be wrapped in a chimera of Indianness with a beat borrowed from Bollywood but with a rap or reggae refrain. Even when Bollywood songs are played in standard Hindi, there are few in the audience who can understand a word of it. Thus these stations represent not India or being Indian, but a particular construction of being Indo-Trinidadian, a set of imaginaries that, collectively, represent an identity.

To be clear, references to collective imaginations of identity are not meant to be disparaging. In fact, the notion that it is somehow wrong to call an identity claim imaginary demonstrates the extent to which we are invested in sacrosanct notions of culture. To question our heritage is somehow automatically an insult, even if that heritage is demonstrably an imaginary set of constructs (which many, in great part, are). In my previous work, I have argued that many of our cherished ideas in many different cultures are either fictitious reconstructions of past ancient events, or modern imaginings of narratives to support whatever our identity claims demand. I continue to hold to that position here.

To further complicate this issue from a cultural standpoint, we are no longer just concerned with the cultural imaginations, ancient or modern, of a small community in a defined area. There are two global dimensions to Trinidad's Indian radio. The historical global dimension brought Indians to Trinidad and Tobago through indentureship. The modern global dimension includes the human-migration dimension that has created secondary diasporas of Indo-Trinbagonians in several metropolitan centres in North America and Europe. This modern global dimension also creates the conditions for local radio stations in Trinidad to have global reach. With the evolution of audio-streaming technologies enabled across the globally connected Internet and accessed through the World Wide Web and on mobile devices, radio stations that would have catered to small localized audiences in Trinidad and Tobago have now become global broadcasters. These stations simultaneously serve their local and foreign audiences, acting as a conduit for greetings and song requests and broadcasting news and current affairs as well as music to listeners in US and European cities, far-flung parts of Asia, and even small communities in other Caribbean territories. To this media-programming dimension they also add social media engagement, enabling audience participation with announcers and hosts through Facebook posts, tweets, and text messages. These stations also create global events, hosting parties and concerts locally and abroad – creating, as it were, cultural events that bolster the collective imaginary of cultural identity and belonging and connect the Indo-Trinidadian diaspora with its ancient and modern "homes".

A Word on Methodology

The present work has relied on many years of research in which the author has approached the phenomenon under consideration in several ways. Bits and pieces of this diverse and messy data set have found its way into the current manuscript. Each approach has formed the basis of some prior research and guided the publication of previous academic papers on this topic.

Statements about the content of the stations considered here are possible because of numerous rounds of content analysis conducted on recordings made at different periods. Some of these recordings were made in the early 2000s, when some of the stations became available through Internet streams. Others were made several years later, in 2013 and 2014, for other papers on the stations. This approach involves coding content along pre-determined or emergent category schemes and using the counts from these coding schemes as the basis for analysis. Such content analysis can be useful for providing descriptive data about media content (such as how much of what kind of song is played or what percentage of time is spent on advertising) and for providing the basis for statistical testing (for example, are there significant differences in the average length of English versus Hindi songs played?).

Content analysis was also used in slightly less conventional ways to gather and evaluate the sentiments of listeners using the Facebook pages of these stations, where available. In these efforts, the author gathered publicly available comments on Facebook to evaluate responses from audience members, with a particular focus on how local listeners differed from foreign or diaspora listeners.

The interview was also a key methodological approach used in the present work. A combination of short, structured and long, semi-structured interviews provided rich data from listeners and from key individuals involved in broadcasting and in various fields related to Indian radio in Trinidad and Tobago. Since this happened over a period of many years, approval was sought and obtained from relevant human subjects or internal review boards where available. Interviewees were duly informed of the nature of the project and other relevant details such as the identity and affiliation of the primary investigator, and their explicit permission to be interviewed and recorded were noted. In most cases interviews were recorded on tape or disk and transcribed for later use. Where possible, interviewees were provided with relevant sections to ensure that their words were used in context and within their understanding of their meanings. Some twenty-five interviews were conducted between 2015 and 2016, while another thirty or so were conducted periodically from 1998 to 2014.

Some of the content presented in the following pages is also the result of documentary research involving both physical and digital assets. While modern digital databases make finding documents much more easy and convenient, searches can still consume hours of time and cross-referencing. The most challenging efforts, however, involved tracking down physical documents and careful examination of aging paper by hand, evaluating relevant content and making manual notes. Newspaper searches were also an important component of the current research. Where these involved digital collections, the ability to reach into history and reconstruct events and situations was truly illuminating. Where the source material was not digitized or indexed, the search was much more tedious and only rarely as productive.

Academics in anthropology, folklore and many other culture-related fields debate the value of what they term emic and etic approaches – research from within and from without. In media studies, there is less of a debate on this, in that much of our work involves studies of content. Where researchers in media studies investigate media organizations and phenomena such as Indo-Trinidadian radio, the emic/etic distinction rarely, if ever, emerges. Media researchers are often both insiders and outsiders to their subject areas. In the present case, the author is a past journalist in Trinidad and Tobago with an insider's understanding of the politics and business of small media operations there. However, as an academic coming from the United States who has not lived in Trinidad and Tobago or worked in its media for more than twenty years, I was able to play the role of the outsider, using what might best be described as a case-study approach in which I used interviews and observation as the basis for data collection.

As a corollary to all of these different investigative techniques, the author was also privileged to benefit from the willingness of many people to contribute information and insights on both a formal and informal basis. These numerous collaborators were often eager to have their say when they learned of the nature of my project.

The Politics of Culture in Trinidad

It is very difficult to discuss culture in Trinidad without treading into politics. This is true of social politics and the dialectics of power relations in day-to-day life, and it is also true of politics in the sense of struggles for control of government. In Trinidad and Tobago, the identity struggles in which cultural voices jostle for position within the emerging schema of national identities have, for the decades since indentureship, presupposed (and manifested in) alignment with racial/ethnic groups. Thus both the formal (that is, governmental) politics and the politics of everyday life are tightly bound to questions of culture and identity, which have in turn been bound (despite generations of mixing, syncretism, hybridity, intermarriage and various other fusions) to essential notions of race and ethnicity.

These essentialist associations are neither academic nor intellectual in nature (though both academics and intellectuals are guilty of them) but rather deeply emotional and tribal in nature. They involve unchanging stereotypes held of either group by the other despite a rapidly evolving society and a flood of countertypes. In any faction of this factious struggle, adherents find their views reasonable and productive, while the views of the other group are dangerous and destructive. This kind of tribal societal division, of course, is not unique.

However, when these divisions become drawn not around personal beliefs or views, but rather broad associations between skin colour, hair type and politics, these assumed connections and the preconceptions that accompany them have the effect of closing the scope of debate. Essentialist stereotypes have the power to "fix" meanings, as Hall (1997) suggests, making it impossible to break out of established moulds of discourse even when they fail to be relevant.

It is in this context that one must tread carefully through the cultural landscape of Trinidad and Tobago. An investigation of Indian radio, for example, might in itself be perceived as a kind of political venture. A reviewer of an early paper on this very topic (co-authored by a Norwegian researcher) argued that the paper "smacked of East-Indian triumphalism" and declared it "revisionist history". The reviewer also referred to Trinidad's Indian music promoters in derogatory terms. While reviewers are all-powerful in the peer-review process, they are also human, and subject to the vagaries of stereotyping and essentialism.

At the same time, it is also easy, in an investigation of a cultural phenomenon, to be swayed by one's own cultural biases – often unconsciously so. In this work, the author acknowledges many such biases, but they are not, perhaps, the ones that might be expected. Key in this discussion, for example, is the idea that the cultural and personal identities expressed in Indian radio or Indo-Trinidadian culture are not in fact "authentic" (a specious concept), "traditional" (having been reinvented many times) or essential. The author argues, instead, that these identities are cobbled together by choice, sanctified by time, consensus (and, often, ignorance) and adopted (or rejected) often as tools of political and social objectives. This is a very different approach from the culture/identity purists who posit the Indo-Trinidadian cultural identity as not only necessary and personally indispensable, but also as an inevitable component of one's social being and a matter of pride and distinction.

As numerous political commentators have noted, ethnicity, race and/or culture in Trinidad and Tobago are associated with politics at various levels not only on the basis of nationalism, ethnic heritage or pride. These associations are also conditional on the notion that social or political dominance is necessary as a means to secure a fair share of scarce national resources. From a traditional view of culture this battle can be seen in the ever-present debate on government subsidies to cultural groups and events, including the traditionally Afro-Trinidadian (a tenuous generalization today) Carnival and comparisons with monies given to Indo-Trinidadian observances. From a broader view, access to resources might be manifest in a particular area's being served through road repairs (this, too, being an outdated reference to traditional geographical segregation of neighbourhoods).

The postcolonial evaluations of Trinidad and Tobago society along such easy lines of demarcation, however, are increasingly being questioned. More current work (Meighoo 2008) suggests, for example, that these simplistic divisions are increasingly tenuous, with shifts of geography, occupation, and other factors redefining and blurring the comfortable boundaries that once clearly separated ethnic groups into categories such as urban and rural, professional and agricultural.

Terminology

In the present work several terms require some interrogation. These terms, their evolutions and their use can often warrant extensive examination and explanation. Since space prevents a thorough explanation of all these problematic terms, it is necessary to select some of the more important and problematic ones for demarcation early in the presentation. The term Indo-Trinidadian, for example, refers to a modern conceptualization of the identity claims of people of Indian ancestry who trace their presence in Trinidad and Tobago to the history of British indentured-labour practices in the nineteenth and twentieth centuries. I emphasize this because there were times in which the notion of an Indo-Trinidadian was itself a contentious issue. As Kris Rampersad (2002) has noted, letter-writers to the newspapers in the 1800s might describe themselves as "A Son of India" or "A Son of the Glorious East", with the use of the sobriquet "An Indo-Trinidadian" emerging in 1888.

While it is an acceptable term, Indo-Trinidadian does in fact inadvertently exclude the island of Tobago from its scope of reference, an issue which has not received tremendous attention because traditionally, that island had so few descendants of the indentured labourers among its population. Despite Rampersad's recording of the term Indo-Trinidadian in 1888, it did not evolve as a preferred nomenclature until well past the 1970s, when the term East Indian might be equally likely to be found, even in government documentation. Where possible, the present work uses the term Indo-Trinidadian.

The notion of "Indo-Caribbean" also comes into some question. As an adjective to describe the larger cultural nexus of Indian heritage in several communities across the Caribbean, such as Trinidad and Tobago and Guyana, this term is a useful one. However, as I will raise in later discussions, there is no such thing as an "Indo-Caribbean", either as an individual or a part of a defined group. The term used as a noun will arise, particularly in third-party quotes, but with some awareness of its weakness. This is a terminological weakness that exists in broader literature about the Caribbean in which some misguided authors refer to the people of the region as "Caribbeans".

Some Notes on "Ethnic Media" and the Politics of Immigrant Identities

Since Indian-format radio in Trinidad and Tobago caters to a specific ethnic audience, it may be considered under the label of "ethnic media". Much of the academic social-science literature has addressed media outside of the commercial mainstream that have a particular cultural or group focus in terms of "ethnic" media or minority-ethnic media. The various explorations of these media have emphasized tropes of difference and questions of ethnic identity, with a general tendency towards treating such media as tools of immigrant cultural maintenance and/or assimilation. As early as the 1920s (prior to the commercial development of radio) scholars were already investigating the roles and functions of media focused on ethnic minorities. Evidence of this may be found in Robert E. Park's publication of *The Immigrant Press and Its Control* (1922). Park, considered by many to be the father of sociology, focused on several issues which persist in studies of ethnic media, including assimilation, nationalism, language and culture. Pretelli (2013, 442–43) noted that in the urban Italian immigrant communities in the United States widely characterized as "Little Italies", these ethnic media were both important and widespread, with over a thousand newspapers in Italian being published in the United States from 1850 to 1930, such that "the ethnic press was pivotal in helping immigrants comprehend (and adapt) to the new land".

In response to negative stereotyping of immigrant communities typical in mainstream media, ethnic media provide the basis for oppositional discourse that not only questions the dominant hegemonies of host societies but also creates opportunities for positive portrayals. For Bratu (2014, 199–200), these media offer migrants "the possibility of self-(re)presentation in the new country", helping migrant communities "gain a voice within the host society".

Under these broad terms scholars have investigated newspapers, radio and television stations that serve or are produced by a variety of ethnic minority groups in many countries. Halter (2013), for example, has outlined the migration of Cape Verde islanders to the United States starting in the latter half of the nineteenth century. Following a period of drought and economic hardship in Cape Verde, many of the inhabitants of what was a Portuguese colony at the time made their way to the New England region of the United States, eventually using mass and digital media to express their cultural heritage.

There exists within these explorations of ethnic media an implicit temporal difficulty that is frequently overlooked. This has to do with the descriptive titles of immigrant, ethnic and minority. When, for example, does a community evolve from being an immigrant community to being an ethnic minority, and when does an ethnic community become accepted into the mainstream?

In Trinidad and Tobago a similar question arises with continuing perceptions of an Indo-Trinidadian community forming the "ethnic minority" despite being (depending on the statistics used) either equal in numbers to the other main ethnic group or in a slight majority. The "ethnic media" frame thus faces some difficulties when applied to Trinidad's Indian radio environment. The clear demarcation between immigrant minority and dominant established majority evident in "ethnic media" scholarship is not as clear in the Trinidadian context, since the community served by Indian radio is numerically approximately equal to those who are not in its target group, and the Indian community is no longer characterized by immigrants, but rather people who have settled for several generations with little or no additional immigration.

Despite the fact that ethnic media are defined in a variety of ways and include a wide range of media products, there remains a tendency to ghettoize these media and, in doing so, exclude them from a role in the mainstream of societies. Ross (2014, 1316) noted that "ethnic media models tend to categorize ethnic media as alternative, diasporic, community or ethnic minority language media". Yu (2015, 133), similarly, wrote that "there seems to be a rigid perception that ethnic media are media only by, for, and about ethnic communities, and any attempt to extend their role to broader society is questioned at best or dismissed entirely as nonsense".

The diversity of form that may be found within ethnic media reflects the diverse strategies that diasporic or multicultural communities may use for exploring transnational and transcultural content. This is true even within a single media platform such as radio. Two examples of Indian ethnic media in the United States reflect something of this diversity.

In the ethnically diverse and well-populated areas of San Francisco, San Jose and Oakland, California, a radio station with the call letters KLOK (otherwise calling itself Desi 1170 AM) has, since 2009, boasted of being the largest and most powerful South Asian radio station in North America (KLOK AM 1170/ Desi 1170 AM, 2016). The station, with a dedicated Indian content format, offers a wide range of programming, including live hosted music, singing competitions and call-in shows (Das 2010) and attracts both listeners and sponsorship from the surrounding urban areas. On the other hand, Venugopal (2001, 20) described a much more modest effort at Indian programming, with *Music of India with Nag Rao* broadcast at the time in "the open wilds of Fairbanks, Alaska, just outside the Arctic Circle . . . a town with fewer than 1,000 Indian families". The latter effort was much more typical for many years, as Indian communities or Indian music enthusiasts found ways to negotiate programme time on community or commercial stations for their specialized broadcasts.

In Trinidad and Tobago, Indian-format radio, which may arguably be considered a form of "ethnic media", raises questions about the role of such media in identity processes, cultural preservation and societal voice. These are among the many questions that we will examine in the following chapters, using the methods outlined above and within several conceptual frameworks including historic globalization, media history, imagined communities and hybridity.

CHAPTER 1

Trinidad and Tobago
Historical-Global Developments

The small Caribbean nation of Trinidad and Tobago has a population estimated at approximately 1.4 million people in 2016 (CSO 2017). The population
comprises several groups whose ancestors came to be there as a result of historical global forces. Native peoples of the region included groups that have been
termed Carib and Arawak. Of the two groups, only a small remnant population of Caribs still exists in Trinidad and Tobago. These groups are thought to
have inhabited the islands of the Caribbean dating back to between 9000 and
8000 BC (Boomert 2016). The experiences of the native peoples after the arrival
of Spanish explorers to the Caribbean were overwhelmingly negative. Between
the late fifteenth and early sixteenth century, Spanish conquistadores wiped out
entire native populations in many of the islands (Lopez 1990; Stannard 1992).
Lopez (1990, 4) summarized some of the observations of the Spanish priest
Bartolomé de las Casas this way:

> One day, in front of Las Casas, the Spanish dismembered, beheaded, or raped 3000
> people. "Such inhumanities and barbarisms were committed in my sight," he says, "as
> no age can parallel. . . ." The Spanish cut off the legs of children who ran from them.
> They poured people full of boiling soap. They made bets as to who, with one sweep of
> his sword, could cut a person in half. They loosed dogs that "devoured an Indian like
> a hog, at first sight, in less than a moment". They used nursing infants for dog food.

As the Spanish decimated local native populations, they also began to settle
the islands and mainland territories of the region, followed closely by competing

European settlers who were willing to do battle to establish and defend agricultural plantations. In order to make these plantations profitable, European powers initiated one of history's most extensive and deplorable human-trafficking schemes, in which African people were taken from their homes and sold into slavery in European colonies, including the United States and islands of the Caribbean. This human-trafficking scheme, with the support of European governments and even many church authorities, started as early as 1501 and would continue until the 1800s. As a result of this scheme, descendants of approximately half of Trinidad and Tobago's citizens were brought to the islands with no possibility of ever leaving. Estimates of the percentage of people of African descent indicate a figure of about 34.2 per cent of the total population (CIA 2016) with an additional 20 per cent or so of mixed (primarily African and Indian) descent (UNFPA 2012).

Approximately 35.4 per cent of the population of Trinidad and Tobago is of Indian descent (CIA 2016) plus the additional 20 per cent of primarily African/Indian mixed population (UNFPA 2012) and small segments of European, Middle Eastern and Chinese heritage. With the distance between the Indian subcontinent and the Caribbean archipelago, a predominance of people of Indian descent would seem an unlikely circumstance but for the role of British global colonial enterprise and its use of slavery as an economic production system.

In 1834 social pressure across Britain precipitated the emancipation of slaves in the colonies (descendants of African slaves shipped there during the previously outlawed slave trade). When these ex-slaves refused to return to the plantations as paid labour, the British had to search for alternative labour sources. Initial experimentation with Portuguese and Chinese labour proved unsuccessful and the British turned to the Indian portion of their empire, with a scheme of indentured labour under which Indian workers would agree to being transported for a limited period of servitude in the Caribbean territories with a promise of eventual payment and return to India.

Trinidad received over 140,000 labourers from India (Look Lai 1993; Roopnarine 2009; Tinker and Birbalsingh 1989; Wilson 2012), while several thousand others went to other Caribbean territories, including, notably, Guyana. Descendants of these indentured labourers now account for significant portions of both these countries' populations. Jain (2004, 8–9) described the indentureship system as it related to Trinidad under the provisions of the British Empire's immigration ordinance of 1854 as follows:

The immigrants signed contracts in India that bound them to certain terms for the period of indenture. On arrival in Trinidad the Indians were assigned to a plantation, where they had to work for three years. Then they had the option of choosing their employer for the following two years to complete the total of five years of mandatory industrial residence, the pre-requisite for becoming legally "free." To qualify for the free return passage to India, the immigrants had to reside in the colony for a total of ten years. After completion of the industrial residence they were again free to choose their occupation and employer for the remaining five years. The law regarding return passage was revised in 1895 and 1898 so that immigrants who arrived in the island after 1895 had to pay a proportion of their return passage. Only 33,294 immigrants (22.47% of the total) eventually opted to return to India; the majority of Indians gradually abandoned the hope of one day returning to India.

For those who did return to India, their fates were sometimes scarcely better than those of their counterparts who remained in the foreign colonies. Two articles from the *Times of India* are instructive in this regard. The *Times of India* ("The Returned Emigrant", 22 October 1932, 10) described the plight of some one thousand "destitute . . . returned Indian emigrants" from the Caribbean "herded together in wretched huts" at a place known as "Matiabruz" in Calcutta and their confrontations with authorities:

> They eke out a mere existence by doing odd jobs when they can get them; some are practically starving. What makes them collect at Matiabruz is their belief that when an emigrant ship comes in they will be able to go back in it; hence their desperate efforts to gain assistance for this purpose from the Protector of Emigrants in Calcutta. They cannot understand why he should not be able to get them aboard, and in their passionate resolve to gain their desire they invaded his office the other day, behaving in such a way as to bring them into collision with the police.

The same *Times of India* article also argued that these events demonstrated that indentured labourers enjoyed a better life in the West Indies than they could make in their native India, but their desire to return to India was based on an abundance of sentiment and longing for their homeland. The paper noted (ibid.), perhaps with some underestimation of the complications of the issue, that most labourers were entitled to free return passage and most who did return enjoyed successful reintegration. This was of little comfort to the stranded returnees, however, since a few days later, on 25 October 1932, frustrations erupted into confrontation once again, as the *Times of India* reported ("Stranded Indian Immigrants in Calcutta", 26 October 1932, 11) that the police had again to be called in "when about 1,000 returned immigrants besieged the office of the Protector of Immigrants in Mission Row, Calcutta, for return to their countries of adoption":

> Many of them returned to India after many years of absence to visit their country homes, and, having been to their villages, found that they were not wanted there and were considered outcast. Many of them had actually set out to see the homes of their parents in the villages. The immigrants, most of them at least, left their village homes where they were unwanted, and journeyed again to Calcutta and finally established a colony at Metiaburuz, Calcutta, living in huts and *bustees* (shanty towns).

For reasons having to do with colonial manipulation of the system of return passage (and inducements of land in lieu of passage) the majority of the indentured labourers to Trinidad remained on the island. As Jain noted (2004, 9) "for many it was not a conscious decision, and only practical circumstances precluded any possibility of return".

Their descendants would come to form about half of the population of the nation today known as the republic of Trinidad and Tobago. This small Caribbean state, with its historically transplanted and culturally diverse population, provides an instructive case of a media phenomenon that incorporates historical global forces, cultural exchange and modern global communications. Indian radio in the two-island state of Trinidad and Tobago (Indian stations exist only on the larger island – Trinidad) is the result of particular historical global forces; its social, ideological and communicative forms and functions are the result of cultural exchange both within and without its local context; global consumption of this terrestrial radio content by way of the Internet marks its transcendence of traditional notions of the local.

Palmer (2003, 496) argued that "the social structures, economic base, and cultural identity of the Caribbean are anchored in its history of slavery, indentureship, and the movement of labour and European interstate rivalry", noting, as well, that eminent Caribbean historians have characterized these phenomena as early forms of globalization. Ho and Nurse (2005, ix) have similarly argued that "Caribbean society and culture are simultaneously diasporic and global", adding that "the modern Caribbean is 'twice diasporized' – both a point of arrival and a point of departure in the long-term processes of globalization and diasporization" characterized as follows: "The modern Caribbean is a point of arrival as a consequence of the spread of European centred capitalism and colonialism, which started some 500 years ago. The region's cultural development has been shaped by the virtual extermination of the indigenous population, the domination of a transplanted European elite, the forced enslavement of Africans, the indentureship of Asians and the integration of other groups from the Middle East."

The term "diaspora" originates with the Greek διασπορά (for "scattering across"). Among its common early recorded uses is in Greek versions of the Bible to refer to groups of people ejected from their primary historical places of residence. Reddy (2015, 12) has argued that traditional usage of the term as associated with the forced migrations of Jewish and African groups does not necessarily fit with the indentureship experience, writing that, in the case of the South Asians dispersed through indentureship, "members of this diaspora are still culturally associated with India, but through generations they have become more localized in terms of ethnicity and nationalism".

Sahoo and Pattanaik (2014, 1) noted that the term "diaspora" has evolved in modern times away from reference primarily to atrocities and displacement of ethnic and religious groups to becoming a broader term for groups and individuals living outside their homelands but maintaining ties in the context of a globalized social, economic and cultural environment: "It has introduced new ways of looking at human migration in the international domain. It no longer implies relations of diaspora as a memory of the loss of resources; rather, it invokes the potential of diasporas for resource mobilization. This expansion of the concept of diaspora has coincided in the last two decades with the phenomena of diaspora–homeland relations being continuously nurtured through cultural, economic and political participation." For Lukose (2007, 409), the term has evoked something closer to a set of culture/identity claims and concepts described as "cultural productions and identity formations of migrant communities that have become important for the larger political, economic, and cultural transformations that mark contemporary globalization". Diaspora in its traditional sense suggested displacement and loss. In its modern senses it can equally suggest new and hybrid identities expressed in evolving dynamic cultural claims and manifestations.

Though independent of the United Kingdom since 1962, Trinidad and Tobago's society and culture include such global influences deeply embedded not just in tradition, but also in implicit ideas, attitudes, social conventions and intergroup suspicions. The society of this two-island state is predicated on global forces that have determined not only the composition of its population but also the character of its culture. These forces and influences are often hidden and powerful, influencing social norms, cultural nuances, and even national policy and politics (Roopnarine 2009; Wilson 2012). Balliger (2006, 278), for example, noted that, in Trinidad and Tobago, "everyday lives are embedded in the cultural legacy of the British Empire", including the use of English, parliamentary politics and the popularity of cricket.

Balliger (ibid.) also emphasized, citing Gupta and Ferguson (1997, 4) that such legacies "must be understood as complex and contingent results of ongoing historical and political processes". These "embedded globalities" arising from a mix of historical experiences (often reinterpreted or reinvented) and ongoing local cultural dynamics are infrequently interrogated or deconstructed – processes made more difficult in the context of modern global influences from the United States and, to a lesser extent, India, Africa and Europe. Wilson (2012, 126), demarcated the cultural mix into two main "folklores", or broad cultural traditions, arguing that "the Creole folklore is a mixture of African, Spanish, French, and English cultures . . . seemingly shared by Blacks, local Whites . . . and the mixed populations" while "the East Indian folklores are those inherited from the indentured workers from Southeast Asia (India and Pakistan), specifically the Hindu and Muslim religions and cultures".

While Wilson's (2012) nomenclature may represent a somewhat simplistic division, it is not altogether inconsistent with general social perceptions in Trinidad and Tobago and the traditions of analysis. Within these traditions it is also *de rigueur* to analyse the levels of fidelity to or maintenance of some notion (however faint) of Indian culture or folklore. Manuel (1997–98, 18–19) described the indentured labourers' ties with India as "steadily fraying" during the period of indentureship and cited a further decline in more recent times:

> The Indo-Caribbeans . . . came mostly as illiterate peasants from an economically-depressed hinterland region of India, with which there has been little contact since. Such contact as has existed has traditionally been limited largely to the goods imported by a few merchants, the occasional visits of Hindu and Muslim proselytists, and, since the 1930s, imported commercial films and records. Even such limited and unidirectional contacts largely involved aspects of the Indian pan-regional "Great Tradition" or commercial mass culture rather than the distinctive regional culture of the ancestral Bhojpuri region. . . . As a result, local concepts of "India", while

remaining important, have become singularly tenuous, idiosyncratic, and in many cases, imaginative.

I will avoid Manuel's (foreign) academic appellation "Indo-Caribbeans", since people of the Caribbean do not generally refer to themselves as "Caribbeans" – though terms such as "Indo-Caribbean" or "Afro-Caribbean" are used to describe various communities. However, more importantly, the "imaginative" component of identity to which Manuel draws attention is one with which the current discussion will engage at several points. Such imaginative identity involves complex processes in which (for Indo-Trinidadians) diluted ancient attitudes and beliefs mix with modern social cues (often mediated today through music, television, films and radio as local or Internet broadcasts) to create a particular set of identity claims. This imagination of identity is a dynamic, constructive and negotiated process in which intersubjective notions of cultural norms and ideals are developed and changed over time, often in response to the historical and social forces (including global forces) influencing the community. It is a process that began during indentureship and continues today.

Roopnarine (2009) has argued that particular dominant global forces surrounding indentureship, including Western capitalism, colonialism and British Victorian attitudes, encouraged Indian migration to the region, and is among those who characterize the several waves of migration as a process over which the Indians had little control in the face of such forces. In the context of the racial and historical politics of the region, this is a contentious point. It is contentious because two narratives divided along ethnic lines have competed for space among Caribbean academics. On the one hand is the view held by Roopnarine (2006, 2009) and others that has emphasized the coercive power of the colonial masters in India and in Trinidad and the manipulation of laws and regulations that ensnared these colonial subjects into a distant indentureship and (later) denied most of them their promised opportunity to return to India. Reddock (2008, 41), for example, noted that the system of transporting and using Indian indentured labour in the Caribbean was characterized in its own time as a "new system of slavery". Wahab (2011, 209) similarly dubbed the phenomenon "a system of quasi-slavery".

There is a countervailing argument that seeks primarily to distinguish the Indo-Caribbean ancestral experience from the broader Caribbean context of slavery. Typical in this regard are the emphases in the work of Fergus (2008,

16) who characterized the indentured labourers as "immigrants" and "settlers", writing, for example, that "the new immigrants were to be free settlers, receiving generous grants of land on which to become independent cultivators of rice and even sugar" and noting that "they were also to be guaranteed the preservation of their family traditions and religious culture".

These differences in perception reflect a larger dichotomy of discourse and competing narratives surrounding race, ethnicity and identity in the region, and particularly in Guyana and Trinidad and Tobago, where Indo-Caribbean and Afro-Caribbean populations contest cultural space. Writing about Trinidad and Tobago in particular, Wilson (2012, 126) argued that the struggle between the two main ethno-cultural groups is partly predicated on "a struggle for cultural recognition and respect" in which the dominance of the Afro-creole folklores has "left Indo-Trinidadians feeling that the national identity of the country marginalizes them even though they account for 40 percent of the population". Beyond the notion of cultural space, debates have also raged for many years over access to jobs in the public sector as well as to the roles of the two main ethnic groups in local media, with Indo-Trinidadians having argued that they have been underrepresented in these areas.

As observers such as Klass (1961) have noted, the development of the petroleum industry in Trinidad and Tobago brought structural changes to the economy and society, partly manifest in the movement of Indo-Trinidadians from traditional agricultural and rural occupations into direct competition with Afro-Trinidadians, who had previously come to be entrenched in public-sector and other mainstream occupations. Thus the competition for cultural space was mirrored by emerging competition for economic space among the main ethnic groups.

These ethnopolitics, while often intense and contentious in Trinidad and Tobago, have rarely spilled over into open confrontation and violence. By comparison, the Guyanese post-independence experience has included devolution from debate and discourse to strife and violence. While Sanders (1978) primarily recounted the history of broadcasting in Guyana, this narrative could not avoid the experience of open racial violence and its roots in the colonial experience. Sanders (ibid.) attributed the hostility between those of Indian and African heritage primarily to the colony administrators during British colonial rule who found it expedient to pit these two groups against each other in the interest of avoiding a unified resistance to colonial exploitation.

Other scholars dealing with the Indian diaspora have avoided the notion of ethnopolitics altogether, arguing instead for a kind of innocent and isolated cultural project devoid of political implications. Jain (2004, 23), for example, distinguished between notions of Indian culture as identity claims and their use as a political tool, even in the context of open ethnic conflict such as that in Fiji, contending that "in the Indian diaspora . . . Indian culture is used not as a means for attaining political power but for identity maintenance and enhancement through the building of temples, preservation and propagation of Indian language, demands for more time for ethnic programs in the electronic media, and so on".

While well-established voting patterns in Trinidad and Tobago and Guyana clearly point to politics of ethnic alignment (in which ethno-cultural markers can be powerful political symbols) and may contradict Jain's perspective, his attention to the role of electronic media is germane to the broader issues at hand. Further, the connections among cultural expression, media and political power are the focus of many other investigations. Balliger (2006, 284–85), for example, asserted that

> Chutney soca music in Trinidad is typically understood as a blending of Indo-Trinidadian chutney and Afro-Trinidadian soca (carnival) music. Chutney music originally emerged from Hindu religious practices, but its contemporary mass form asserts Indian cultural presence and power in Trinidad. The popularity of chutney soca was facilitated by the launching of the first nationally broadcast "Indian" radio station in 1993, which played a significant role in the election of the first "Indian government" in 1995.

Language and the Indentured Labourers

Nicholas (2009) described the cultural attachments and involvements of young people among a Native American Hopi community with a particular focus on those who did not speak the traditional and actively spoken Hopi language, Hopilavayi. These youth insisted that their connections to their culture were strong despite their inability to speak the language. Linguistic traditions, even those that are vestigial or remnants, can thus provide important anchors for ethnic belonging and identity. Where a group identity exists or is sought, the members of the group may argue that their imagined or real linguistic heritage is threatened or under attack and in need of revival (Eisenlohr 2006; Urla 1993). Eisenlohr (2006, 26) noted, by way of example, that in Mauritius, despite the widespread use of Mauritian Creole, "ideas about ancestral languages play a crucial role for creating communities among Indo-Mauritians, in particular Hindus".

The vast majority of content featured on Indian radio in Trinidad and Tobago consists of songs in either standard Hindi or the local Bhojpuri dialect. In a sample of content these occupied in excess of 80 per cent of broadcast time (Mohammed and Thombre 2014). Outside of these songs, other content, including announcer segments, informational programming, news and audience interactions are all conducted in standard or, somewhat less frequently, Trinidadian English (variously described as a dialect or pidgin form of English). Many stations also include songs that are in standard or Trinidad English in their playlists, as well as songs that include both Hindi and English versions (sometimes as "mixes" or "mash-ups").

The main exception to the English programming norm comes in the form of religious broadcasts that include segments such as readings from holy Sanskrit texts such as the *Ramayana* and the *Bhagavad Gita*. These readings, however, whether in-studio, or carried as live broadcasts from ceremonial observance sites, are almost invariably accompanied by translations.

The linguistic dynamics of Indian radio in Trinidad reflect the tensions between the perceived importance of Hindi as a marker of "Indian culture" and the reality that most Indo-Trinidadians today neither speak nor understand Hindi or the local Indian-origin dialect often described as Bhojpuri. Also called "plantation Hindustani", this latter dialect, which emerged out of negotiated meanings and expressions among diverse language forms brought to the colonies from various parts of India, has been the subject of some linguistic investigation, but barely survives as a spoken language.

Among diaspora communities, the preservation and persistence of ancestral languages is known to vary among different groups and contexts – though gradual declines and adoption of host-nation languages are usually inevitable. Jayaram (2000, 41), for example, noted that the state of language persistence and morphology in various Indian diaspora communities varies tremendously along with the variety of experiences that these communities have endured. Jayaram (ibid.) also specifically pointed to the fact that "in Guyana and Trinidad, *Bhojpuri* is used in folk songs and Standard Hindi in religious services and ceremonies", while English is used in everyday life. Both the distinctive uses of Hindi and Bhojpuri and the dominance of English are related to the historical and political evolution of the Indo-Trinidadian diaspora. British missionary W.H. Gamble (1866, 33) suggested that the linguistic differences among the indentured labourers were so great that English had to become their medium of communication.

Mahabir (1999, 13) detailed the linguistic milieu of the indentured labourers who journeyed to Trinidad ("mainly from the eastern districts of Uttar Pradesh and Western Bihar") as including "various dialects of Indian *Bhojpuri* and other minor languages like Assamese, Bengali, Nepali, Bihari, Punjabi, Oriya, Marathi, Malayam and other tribal languages", while small groups spoke Awadhi, Magahi, Maithili, Tamil and Telugu, and "the size of each linguistic group ranged from several dozens to a few thousands".

The demands of inter-group communication and the legacies of religious teachings resulted in several key linguistic developments that inform the present discussion of Indian radio in Trinidad and Tobago. Since the indentured labourers needed to communicate with each other, they evolved a version of Bhojpuri that incorporated influences from different strains of that language into their plantation Hindustani. Mahabir (1999, 14) described this linguistic form as an inter-group survival language, though he noted that, as Ramesar Mohan (1978) previously argued, this ad hoc, cobbled-together language evolved into the lingua franca of the community for some time.

Mahabir (1999, 17) has noted that "in an effort to communicate with some of the Indian workers, the European estate officials sometimes found it necessary to adapt their own use of English". Mahabir (1985, 1999) also posits that a certain amount of linguistic exchange took place between the Indian indentured population and the Afro-creole population that included the indentured labourers learning the French-based patois of the Afro-creoles and some Afro-creoles learning the prevailing version of Bhojpuri.

For their religious rituals, the indentured labourers also kept alive several other linguistic traditions. It is clear that Hindi, for example, was important for the reading of sacred texts and the conduct of religious rituals among the Hindus who formed the great majority of the labourers (Jayaram 2000; Mahabir 1999; Morton 1916; Ramesar Mohan 1978). Scholarly inquiry has paid somewhat less attention, however, to other languages that survived through religious scriptures and teachings, primarily Arabic and Urdu, within the small

Muslim community (Kassim 2002). It was this diversity of language practices that would become a factor in the initial presence of Indian content on Trinidad radio.

The evolution (or more properly, devolution) of these language traditions involves several factors. The usual attrition of traditional languages over generations was one of these factors. Jayaram (2000, 45) noted, for example, that by the early 1980s the language known as Trinidad Bhojpuri (often casually called "Hindi" locally) could only be found in use among 3 per cent of the population, with many of those being over seventy-five years old and restricted to rural communities. The figure also included some in the age group fifty-five to seventy-five who were described as semi-speakers. This kind of remnant survival of the language tradition was also evident in some elderly speakers of Tamil and Urdu and has likely experienced further declines.

This prior research, as well as the input of collaborators in the present research, suggests strongly that, while rare today, these traditional language practices were also increasingly conditioned towards exclusion through social and cultural factors prior to their decline. One of the very common observations about the use of Bhojpuri, for example, is that sometime before the 1960s and 1970s, it became common for elderly users of the linguistic form (most of whom were by that time also proficient in English) to speak in Bhojpuri among themselves to exclude younger generations who had little or no understanding of anything other than English. Additionally, speaking any variant of Bhojpuri or Hindi in public – particularly in the presence of non-Indians (especially post-Independence) – quickly became an invitation to derision, with mainstream speakers of English mocking the use of traditional terms or language as backward or uneducated (Samaroo 1996), in keeping with what Eisenlohr (2006, 28) termed "a hegemonic national identity tending towards their exclusion". This assertion also fits neatly with Jayaram's (2000, 47–48) observation that "the life of the immigrant Indians was witnessed by the patterns of their ancestral culture being ignored, ridiculed or suppressed by the carriers of the dominant culture of the colony . . . there was continuous pressure on them to Creolize".

Ironically, it was standard Hindi that also featured prominently in attempts to culturally condition the indentured labourers and their descendants not just into creole society but into a foreign set of religious values intended (in part) to bring them closer to the dominant norms of their new colonial home. On this point, Prorok (1997, 380) argued that, as far as the authorities were concerned, "it was in the plantocracy's best interest to Christianize their workforce".

Jayaram (2000), Mahabir (1999) and Prorok (1997) are among those who have detailed the attempts of the Canadian Presbyterian missionary the Reverend John Morton to use standard Hindi as a means of reaching the indentured immigrants in order to convert them to Christianity, and connect those efforts to the spread of standard Hindi in Trinidad. The memoirs and writings of Morton (1916) indicate that the missionary first visited Trinidad on a winter trip to the West Indies, where the mild weather was recommended for his recuperation from a bout of diphtheria. In his wanderings during that early visit he happened upon the indentured Indian labourers, and later returned to Trinidad with his wife and child in 1868 as a missionary. Morton (ibid., 8) argued that the indentured labourers were left without attention to their spiritual lives and that whether they were returning to India or staying in Trinidad, the mission of conversion would be well expended upon them, noting that it was important that they be "saved from their dark idolatry or Mohammedan delusion".

Prorok (1997), however, recounted Morton's establishment of the Canadian Mission at Iere Village in South-Central Trinidad with some reservations about Morton's acceptance into the Indo-Trinidadian community when John and Sarah Morton arrived in 1868, since the indentured labourers did not trust Morton. Following on this notion of distrust, Samaroo (1996, 24) has explored several factors that may have accounted for the Indians' reluctance to attend the government's mixed-race ("ward") schools, including the fear of conversion, noting that "by 1868 there were no more than three Hindu children in the colony's ward schools", and arguing that they resisted from fear of conversion, among other factors.

If colonial authorities could not secure the integration of the transplanted Indians, they sought, at least, to ensure their compliance with colonial norms, including the adoption of English and Christianity. In this regard, Morton and his mission played an important role (whether deliberately or not) in the colonial programme, serving the interests of the planter class (with whose members he was known to be friendly) and the colonial authorities.

Morton's diary indicated that he familiarized himself with the spoken "Hindi" and even the Urdu used by the indentured Indians in Trinidad and determined that it would be through Hindi that he would most effectively reach this population. Additionally, Morton's diaries indicated that as he learned more of the local language he would meet with the indentured labourers in their homes or even on the roadside. Less commonly emphasized is the fact that Morton also communicated with the locals in their emerging English idiom. During one of his first encounters on a sugar estate, Morton (1916, 22) recorded his interaction with some of the Indian immigrants (whom he perhaps erroneously perceived to be priests, since he calls them Babujee – a generic honorific which he translated as "priest") as follows:

> Jan. 20th. – Visited Union Ball and Les Efforts with Mr. Lambert. At Les Efforts fell in with two Babujees, one a fine looking Brahman about twenty years of age and only nine months in the Island. Men of all castes crowded around us. One boasted that he ate beef and pork and everything, on the principle that God made all – beef and rice and rum. My teetotal friend playfully told him: "No. – Devil make 'em rum." His ready answer was: "Then, I devil's man." One of the Babujees argued against eating beef in this style: "When I little 'picknie'" mumma give me milk. I grow big – so, – [with gesture indicating height} – cow give me milk; no kill and eat mumma; no kill and eat cow."

Morton's account bears some question, since it would not be expected that either a Brahmin or any Hindu cleric would boast of eating beef, nor would any other Hindu boast of this to a Hindu cleric. Aside from this, it is interesting to note in this exchange that even recent immigrants were already engaged in using English in some form. This introduces some question about Morton's boasts regarding his acquisition of linguistic proficiency. Indeed, the diary entries and narratives of his wife, Sarah E. Morton, indicate that she encountered a variety of linguistic traditions and communicated with the household help (both Indian and creole) in the local variant of English where possible.

Beyond religious conversion, Morton also engaged with issues of agriculture and welfare and developed a reputation for being a source of help for those in need. Bearing in mind Samaroo's (1996) contention that the missionaries and the plantocracy shared an interest in maintaining a docile labour force in which conversion was a tool of social control, the Indian indentured labourers looked upon the collaboration (or, for some, collusion) between Morton and the white plantation owners as a threat. Collaborators in the current research recalled stories, popular within the Indo-Trinidadian community, that, beyond the many approaches recorded in his diaries by which Morton would reach the Indian indentured labourers, he was also known to employ heavy-handed techniques such as conspiring with plantation owners to have labourers' wages

withheld until the labourers agreed to meet with him and listen to his teachings. One collaborator even repeated an allegation that Morton and his friends
were in the habit of coercing women from the villages to serve in the homes
of wealthy planters when needed. Whether true or not, these stories and even
more serious accusations are still repeated among older members of the Indo-
Trinidadian community with regard to Morton.

In 1872 Morton's mission imported to Trinidad (from India) Christian literature including Bibles and catechisms in Hindi (Jayaram 2000; Morton 1916).
Beyond this, Morton also (with the assistance of locals) prepared collections of
hymns, written in English-transliterated Hindi and (as remembered by some
of the older collaborators in the present fieldwork) set to the tunes of traditional Hindi songs that were well known among the Indians. This Hindi-based
proselytization was evident outside of the formal religious context, since the
Presbyterian mission also set up and ran numerous schools that were designed
to be both culturally and linguistically friendly to the children of the indentured labourers, providing an opportunity for education in a situation where
parents were hesitant to send their children to school with the creole Afro-
Trinidadians. The primary schools were often used as sites for religious services
and Sunday school and, eventually, the Presbyterian schools (sometimes in the
same or contiguous compounds as their churches) began to be synonymous
with the church, and even the notion of Presbyterianism began to be conflated
with being Indo-Trinidadian.

Despite the linguistic enterprises of Morton and the Presbyterians, both the
Bhojpuri and Standard Hindi forms were eroded by the middle of the twentieth
century, with Jayaram (2000, 47) noting that "by all accounts . . . the spontaneously evolved Trinidad Bhojpuri and the deliberately developed Standard
Hindi did not survive for more than a century in Trinidad" and adding:

> If at all they are still spoken fluently, it is almost exclusively by the very elderly in
> rural areas, and as such there is no speech community of these ethnic languages left
> among the Indo Trinidadians anymore. Whatever has survived is mostly in their folk
> songs and in their lexicon of kitchen and food, and to some extent in their kinship
> terminology. The only systematic ethnic use of Hindi in contemporary Trinidad is
> to be found in the religious realm among the Hindus.

The linguistic and religious impacts of the Presbyterian mission in Trinidad
also resulted in broader social changes reflecting an adoption of Western Chris-

tian values, manners and affectations as a gateway to opportunity in (if not integration with) the broader mainstream of British colonial/creole and later independent post-colonial society. As Gooptar (2013b, 146) has noted, "after 1900, with the growing pressures of an expanding society and increasing pressures to prepare children for civil society jobs, many East Indians accepted Presbyterianism as a means to an end", adding that "as late as the 1960s East Indians were still being pressured into changing their names and religion to obtain teaching jobs".

Religion and Culture in Trinidad and Tobago

Two primary categories of religious imports came with the Indian labourers, one broadly termed Hinduism (a problematic term) and the other (about a 15 per cent minority) the Indian version of Islam. For some, such as D. Parsuram Maharaj ("Back to Hinduism in Trinidad", *News India–Times*, 30 October 1998, n.p.), the Hindu community was an obvious and easy target, subject not just to the colonial authorities' Christian biases but also to economic coercion.

The history of competing dynamics of religious retention on the one hand and erosion through Christianizing missions on the other continues to play out in the modern experience as well. Over more recent years the driving force behind missionary efforts has not been Canadian Presbyterians but American evangelicals, first manifest as a broad "Pentecostal" movement in the 1970s, which has more recently been supplanted by various evangelical or born-again churches, which, according to Maharaj (ibid.), "offered social services to the Hindus and sold salvation cheaply to all newcomers".

For several decades (and to some extent continuing today) these evangelical movements worked through local proselytizers, who received money and donations of literature and equipment from foreign sources. Typically, the local operatives (many of whom were of Indian descent) would pitch a tent in an open field in a village and announce the start of a "crusade" for several weeks – the hostility and violence inherent in the term "crusade" apparently being lost on the target audiences. Similar to what is termed a "revival" in other places, these "crusades" featured loud music and singing (notably, in English, as opposed to the Hindi or Urdu that was usually found in temples and mosques) that drew audiences including many younger members of Hindu (and to a smaller extent, Muslim) families. Ramoutar (1990, 1) argued that these evangelical offensives were damaging to Hindu families who were "tired of seeing their sons and daughters suddenly come home wearing crosses and spouting anti-Hindu slogans – tearing up the family in the process".

Maharaj (1998), Ramoutar (1990) and Munasinghe (2001) have all noted that, at least partly in response to both historic and contemporary efforts at conversion, the Hindu community in Trinidad has, more recently, engaged in its own revival efforts, including attempts to revive the use of Hindi even in limited ritual practices. Munasinghe (ibid.) described the development of Hindu youth organizations as one reaction to the Christianizing crusades and to more general cultural drift. Such organizations, according to Munasinghe (ibid., 110), "seek to combat the youths' sense of alienation by relating Hindu precepts to the Trinidadian context", using language and religious instruction and promoting both Hindi language instruction and English translations of religious texts.

On closer examination, one finds some debate as to the very notion of "Hinduism" so casually referenced today. Hinduism as a homogenous, unified, codified and agreed-upon belief system common to Indians across space and time is, arguably, a misnomer, as it has not necessarily existed in that absolute sense for any appreciable time (and may still not exist today). Yet there exist several religious ideals and conventions that followers of Hinduism have held to constitute their religion for many centuries. Debate arises over the degree to which the notion of Hinduism properly unifies diverse, regional variations in belief and practice as well as over the mechanisms of that supposed unification (King 1999; Lorenzen 1999).

One argument in this debate is that the broad-brush notion of Hinduism was a convenient label applied by British colonial authorities and missionaries to a broad range of religious and social practices often unintelligible to outsiders, who made also little effort to understand them and sought primarily to denigrate them as heathen and evil. As European scholars "imposed their own definitions of religion onto a scattered and fragmentary body of indigenous beliefs and practices" (Haan 2005, 14) they came to define these wide belief structures as something they called "Hinduism".

The opposing view, often expressed among Hindu nationalists, emphasizes, instead, the common core of values and ideas that served as a unified basis for religion in India that existed for centuries before British colonial rule and survives today. The debate raises many questions about the presumed uniformity of belief to which modern diasporic Hindus such as those in Trinidad may refer in their construction of their religious identities. The debate on Hinduism as a diverse set of beliefs and practices also informs a more nuanced understanding of the complexity of early indentured society, with its multiple adherents of diverse Hindu beliefs as well as its smaller contingent of Muslims.

Lal (1998, 230) has described the reconstitution of Indian culture and religion among Indian indentured labourers in Trinidad and elsewhere as a set of diverse processes in which "the bits and pieces which survived migration and indenture were knitted into a new pattern to suit the requirements of the new environment". Sherry-Ann Singh (2005, 354) has similarly argued that despite the persistence and preservation of their social and cultural practices, the Indian indentured labourers experienced upheavals that influenced the forms and sometimes the foundations of their observances – even affecting their religious ideas, noting that "the uprooting from the Indian context necessitated attempts at community and religious reconstruction" such that "in Trinidad, elements of religion were variously truncated, modified, diluted, intensified or excised". Singh (ibid., 354) also invoked Haraksingh's (1985, 163) perspective, suggesting that the adjustment processes involved adaptation rather than simple transplanting. Singh (2005, 354) described a period following the abolition of indentured immigration in 1917 when the Hindu community focused on "religious reconstruction", including a "move towards homogenization" of several threads of Hindu belief and practice that had come to Trinidad from various parts of India: "Thus the Hindu community attempted to "reconstruct" in some standardized form, a religious practice that was itself diverse and scattered in its original form, drawing on a wide range of regional practices, mythologies and beliefs from a deep and dispersed Indian tradition." For Singh (ibid.), the move towards homogenization in the early twentieth century accompanied the evolution of particular observances and emphases that defined the local practice, including "the sanctifying and inclusion of many local elements into the Hindu religious realm, the increase in collective religious activity and observances, the proliferation of temples, and the prominence of the *Ramcharitmanas* (a particular written version of the *Ramayana* story) as the most popular Hindu religious text".

Culture and Exclusion in Trinidad

Despite clear evidence that Indian religious traditions were exoticized, othered, denigrated and distrusted, some, such as Sherry-Ann Singh (2005, 355), have argued that Indian cultural vestiges in their various Indo-Trinidadian forms have, over time, indeed negotiated spaces of belonging within the Trinidad and Tobago national identity complex, breaking through boundaries and xenophobia, and that "various dimensions of Indian culture have now been integrated into Trinidad and Tobago's multicultural prism" and attesting to the "nationalization" and "naturalization" of Indian culture in Trinidad and Tobago. As evidence of this Singh (ibid.) included the following observations (including the role of "Indian media"):

> The State now recognizes the validity of Hindu and Muslim marriages, and the Hindu mode of cremation. Non-Christian festivals, rituals and observances are now met with a higher degree of respect and many are accorded adequate attention by the media. *Divali*, *Eid-ul-Fitr* and (Indian) Arrival Day have acquired the status of public holidays. Various dimensions of Indian music, dance and drama can be labelled as essentially 'Indo-Trinidadian'.
>
> The rapid growth of an 'Indian media' has contributed immensely to the publicizing of many dimensions of Indian culture. Indian cuisine has almost totally eclipsed cultural and ethnic barriers.

Many others (Gooptar 2013b; Moore 1995; Roopnarine 2006) point, instead, to the historical and continuing exclusion of Indian culture as an acceptable part of the mainstream of popular culture in Trinidad. Therefore, in the particular argument presented above, Singh (2005) may use legal achievements and niche media institutions to overstate the case for the integration of Indian cultural traditions into the national cultural mix, particularly in the broader context of the arrival and presence of Indian culture sometimes being seen as disruptive.

Quraishi (2015, 409) has argued that global historical forces (under which heading we may include the attempts at Christian conversion and idealization of Christian religion in the colonial programme) played a part in the development of "East Indian" (what we have termed "Indo-Trinidadian") identity, noting that the heterogeneous groups of immigrants from different regions of India with differing languages and practices were uniformly "marked as distinctly inferior as a result of their indentured status and non-Christian religions". Quraishi (ibid.) has also recognized the tensions between the competing nationalisms anchored, on the one hand, in the new homeland and, on the other, in construals of an India that was ancestral in its influence but current as a focus of belonging. This became particularly important as India threw off the yoke of colonial British rule, and nationalist fervour spread to the Indo-Trinidadians who by 1947 had already been Trinidadian for several generations. The notion of being Trinidadian was itself an amorphous concept, since the island was a colony of the United Kingdom with various groups of labour having been imported as its main populations. With the lure of Indian nationalism among a community who did not necessarily always feel welcome in their new home, there was also an association with Hindu identity – construed in a manner that suggested a unified and uniform religious code that denoted common identity markers. This may have been radically different from the fragmented set of regional beliefs that British ignorance chose to classify as a single religion.

Meanwhile, back in India, by the 1950s, media reports touted the "growing importance" of these immigrants to Trinidad and British Guiana.

> Indians of Trinidad and British Guiana, through their rapid growth in population and continued westernization, are gaining a steady importance in the affairs of those colonies. Some informed sources here feel the Indians may all but run things within the next fifteen years. . . . Many of them have become successful shopkeepers, and the money-lending business is predominantly Indian there are as many Indians practicing law and medicine and the Indians studying for those fields outnumber all others two to one. ("Trinidad and British Guiana: Growing Importance of India", Associated Press and *Times of India* News Service, 13 July 1957, 6)

Old and New (Primary and Secondary) Diaspora Populations

Returning to Ho and Nurse's (2005, ix) notion of a "twice diasporized" Caribbean, the migrations of populations of Indian descent both to and (later) from the Caribbean constitute processes of globalization in which historical global processes and modern migratory trends and influences (including globalized communication technologies) have resulted in the development and maintenance of what have been termed "transnational" families and communities (Madianou and Miller 2013; Sutton 1987, 2004).

With specific reference to the Indo-Caribbean communities, Ramnarine (2011, 143) distinguished between the initial displaced populations and their descendants in the adopted home countries (for example, Trinidad, Guyana, Fiji) as the "old Indian diaspora" (or what we might term a "primary diaspora") and their later descendants, many of whom have voluntarily migrated to North American and European metropolitan centres ("secondary diaspora"). The present choice of the terms "primary" and "secondary" helps to distinguish the present focus on successive generational phases of migration in which the primary migrants, originating in India, settled in the Caribbean, only to have their descendants later migrate to metropolitan centres of the United States and Europe. This is necessary since others such as Jain (2004) tend to use "old" diaspora to refer to the historical migrations of indentureship and "new" to refer to the modern migrations of workers directly from India to places like the United States, thereby ignoring (or at least failing properly to contextualize) the intermediate migratory phase to other places such as the Caribbean and Fiji.

With reference to the primary diaspora, even modern migrations, however, are connected to historical forces and the social dynamics imposed by colonial policies and practices and perpetuated through the social legacies of the plantation economy (Beckford 1972) and its structures. Teelucksingh (2011, 140) is among those, for example, who have connected modern-era migration to race relations in the Trinidadian primary diaspora, arguing that race relations were "in a deplorable state" in the 1980s and 1990s, such that Indo-Trinidadians felt relegated to second-class citizenship. Teelucksingh (ibid., 140–41) further argued that the election victory of the predominantly Indo-Trinidadian United National Congress in 1995 and the ascension of Basdeo Panday as the first Indo-Trinidadian prime minister of Trinidad and Tobago did not alleviate the out-migration pressures, since the predominantly Afro-Trinidadian People's National Movement (PNM) returned to power in the next election, coinciding with an increase in violent crime that appeared to target Indo-Trinidadians, resulting in Indo-Trinidadian families selling their homes and businesses and moving to the United Kingdom and North America.

Despite some contention about the accuracy of the assertion that violent crime was used as a weapon against the Indo-Trinidadian community, Mahabir (2005) argued that statistical analysis bears out the perception that Indo-Trinidadians were disproportionately targeted in these crimes, and Prentice (2012) has documented the use of kidnapping as a social and political threat directed at Indo-Trinidadians. Whether these crimes were indeed primary motivations for migration remains difficult to establish, since migration to metropolitan centres from the Caribbean has been predicated on many other factors, including economics and family ties (S.N. Mohammed 1998). Whatever the reasons, large communities of Indo-Trinidadians (and Indo-Guyanese) have made their homes in the United States, Canada and (to a lesser extent) the United Kingdom and are engaged in competing and fluid dynamics of cultural processes which are partly manifest in their media practices.

Ramnarine (2011, 143) pointed to the relationship between the embedded global histories of these diasporic groups and modern technologies, noting that "contemporary creative practices circulate through diasporic networks" and "through film industries, internet and media technologies, festivals, cultural tours and creative exchanges". Ramnarine (ibid.) sought to connect the construction of identity with global contemporary cultural practices and exchanges, while asserting that these are interwoven with the global historical realities of these primary and secondary diasporas. In doing so, she subverted a common and implicit assumption of novelty that pervades many investigations of global media forces. Ramnarine departed from the conventional tendency to assume that global currents of culture and communication are necessarily, exclusively, or even primarily the result of modern globally connected technologies associated with the Internet. Such currents of culture and communication, considered in the *longue durée*, may well be continuations of global avenues of communication as ancient as written letters and proclamations that spread messages to distant lands. They may also be extensions of less ancient global avenues of communication, such as shortwave radio, which spanned the globe even in the early decades of the twentieth century, bringing broadcasts from India to listeners in Trinidad and British Guiana to populations whose belonging to their new homelands was by no means a settled matter and who still harboured very strong sentiments of their Indian identity.

The cultural influence of migration and pre-Internet mass-media forces have also frequently been implicated in the process of identity negotiation among both the new and old Indian diaspora. Sanders (1978, 10) alluded to these forces as they related to the formation of Guyanese national identity, contending that much more than domestic efforts at national cohesion, "the influence of the more developed societies, particularly that of the USA and Britain, has been a far greater unifying force culturally on young people in both groups, as they became less African and East Indian and more 'westernized' in their tastes for music and dress-styles and in their attitudes to religion and other traditional beliefs".

Within this discourse of identity processes, Ramnarine's (2011) notion of the primary and secondary diasporas also alerts us to the fluidity of cultural belonging and the extent to which notions of nationhood and identity have changed in the global course of human events, and how they have changed quite rapidly in the experience of the groups undergoing the displacement, resettlement and redefinition of self and community. This is one of several sign-posts that strongly suggest the need to re-examine the assumptions surrounding culture and identity which have so limited modern academic discourse and, to some extent, even hampered the possibilities of human potential by prescribing (often erroneous or conceptually limited) classifications of self and others.

Conclusion

Indian radio in Trinidad has prompted discourse and debate on how cultural markers are used to assert identities. Thus Bollywood songs that originate in another continent may be invoked as part of the identity of the Indo-Trinidadian without regard to the fact that these songs are in a language incomprehensible to most of the Indo-Trinidadian audience, and that the songs themselves are of questionable relevance to Indian culture – often regarded as poor commercial enterprises that run counter to Indian values. Chutney music, more closely rooted in local experiences and the hybridity of Trinidadian experience (Diethrich 2004; Aisha Khan 2009; Munasinghe 2001), is also cited as an overt identity marker, but one which also needs to revert to English to be understood and one which is widely condemned as being detrimental to the image and ethos of the Indo-Trinidadian community (K. Mohammed 2011; Parasram 2008).

More recent explorations of Trinidad culture have suggested that the traditional strictures of what constitutes culture and what defines a group culturally are being questioned. Indian radio is at the centre of these redefinitions, with programming that is in some cases becoming increasingly inclusive and in some cases positively global. On several of these stations an overwhelming Indian focus is carried through some diverse vehicles such as reggae and soca remixes of Indian songs. As Balliger (2006, 285) pointedly illustrated, this particular global/regional influence may well be a part of the redefinition of traditional cultural boundaries in Trinidad "profoundly transforming culture and political configurations among 'East' and 'West' Indians in Trinidad" as Jamaican dancehall permeated both soca and chutney soca genres.

It would be unreasonable to expect various cultural groups or even concerned academics in Trinidad and Tobago to answer conclusively the question of "what is culture" and closely related questions such as "what constitutes identity". However, the broader cultural boundaries of Trinidad and Tobago within which the specific phenomenon of Trinidad Indian radio operates present several unique opportunities to examine these questions in their historical context, in the context of global interplay and in the context of communications technologies that enable these various influences to intermingle and create.

CHAPTER 2

Culture and Globalization

Several important concepts surround culture and global forces as they relate to the Indo-Trinidadian experience and as a background to the development of cultural expressions such as Indian-format radio. The concept of "culture" itself will be examined here with particular attention to the debate over fixedness versus choice in cultural identity and expression. Culture is chosen as the focus of attention (over other commonly debated social factors in Trinidad, including race and ethnicity) as it provides a broader and more productive basis for understanding social phenomena while including consideration of other dimensions of national identity.

The discourse surrounding Indian-format radio in Trinidad and Tobago inevitably makes reference to the concept of culture in its many and varied dimensions, including tradition, identity and even social power and national integration. Landgraf (2005, 26) considered it important that we "understand the modern interest in culture – and subsequently the multiplicity of congruent and incongruent meanings of culture – as a cultural phenomenon, as a defining aspect of modernity's own culture". Appreciation of this reflexive position provides a starting point for divesting "culture" from absolutist and essentialist approaches embodied in traditions of racial demarcation, national stereotyping and broad sweeping generalizations about the tendencies and values of huge swaths of humanity.

In the context of Trinidad and Tobago, the very conceptualization of culture may also be seen as a function of history and politics in which overt cultural expressions such as music, religious and social events can be a part of the struggle for social power which is also inextricably linked to overt racial and ethnic markers. A revised, broader, or more reflexive view of culture, however, allows for emerging notions of modern national ideas and expressions to be considered outside of the strictures of race, ethnicity and the dictates of compulsory identities so long entrenched by colonial experience and embedded in local politics and discourse.

Despite the existence of many volumes of respected scientific inquiry on the topic, no single and accepted definition of "culture" exists. Kroeber and Kluckhohn (1952) compiled more than 150 definitions of the term spread over several categories and domains of knowledge including diverse areas such as genetics and psychology. In addressing what he termed the "banalization of the concept of culture", Lasky (2002, 72–73) noted that the term itself has given up much of whatever meaning or meanings it may have once held, if only through its overuse in diverse contexts.

Lasky (ibid., 73) noted that the term "culture" may today be found in an Englishman's call to save Marks and Spencer because it was part of British culture, in corporate criticism of aircraft manufacturer Boeing for lacking "a culture of hire and fire" or in calls for a "cultural revolution" inside the Central Intelligence Agency and the Federal Bureau of Investigation after 9/11, concluding that the word "culture" is "an all-purpose semantic filler wadded into any hole in an explanation".

One of the earliest and most widely used definitions of the term "culture" is taken from Sir Edward Tylor (1871, 1) who described culture as follows:

> Culture or Civilization, taken in its wide ethnographic sense, is that complex whole which includes knowledge, belief, art, morals, law, custom, and any other capabilities and habits acquired by man as a member of society. The condition of culture among the various societies of mankind, in so far as it is capable of being investigated on general principles, is a subject apt for the study of laws of human thought and action.

Hutnyk (2006, 352) noted that this definition was not one of "levelling egalitarianism" but rather a condescension in which the nineteenth-century anthropologists viewed cultural development as ranging from "the savage to the civilized" and within which view they were loath even to encounter the savages and heathens. This particular incarnation of the culture concept concerned itself not so much with commonalities of belief and behaviour among groups but more with questions of refinement and achievement of distinctions or graces by a privileged elite, as well as the amassing of works of fine art in society. Taken as a whole, Tylor's *Primitive Culture* is consistent with the perspective presenting culture as a state of refinement defined in various levels reflecting greater or lesser degrees of civilization and intrinsically connected to particular races of people.

Benhabib (2002) suggested that the dominant conception of culture remains the egalitarian approach of the likes of Bronislaw Malinowski, Margaret Mead and Claude Lévi-Strauss, emphasizing the ubiquity and universality of cultural forms as well as the social construction of meaning in specific contexts. However, Benhabib noted that the attendant assumptions around culture stemming from these early anthropologists take for granted that all human groups possess some kind of culture, each distinct and definable. Mathews (2000, 3) had similar concerns, writing that "the assumption common to these writers is that there are discrete patterns of cognition, values, and behaviour that members of each of these groups share in common, in contrast to members of other groups". Benhabib (2002, 4) suggested that the underlying premises of this "reductionist sociology of culture" are faulty in their assumptions that (1) "cultures are clearly delineable wholes", (2) cultures are congruent with particular population (some others would say "ethnic" or geographic) groups and that (3) diversity within cultures presents no problems for policy.

Douillet (2005, 22–23) pointed to this very failing in analyses of Caribbean culture, including Trinidad's, writing that early scholars of Caribbean society and culture were more interested in the retention of cultural traits among "African" or "Asian" groups in the region and establishing "cultural catalogues" of specific groups, failing, in the process, "to take into consideration the complex sociocultural effects and processes of cultural interaction between groups". To these interactions may be added the modern intricacies and historical complications of global influences, some (like the Hindu and Muslim traditions that travelled from India with the indentured labourers) fixed and remembered, some current and contestable (like the past and present influx of Indian films and film songs) and others imagined and transformed (like syncretic inventions such as chutney soca).

Aisha Khan (2009, 165), however, noted something of a shift away from the perspectives of fixedness that elicited concern from Benhabib, Mathews and Douillet toward evolving notions of cultural fluidity and its various manifestations, writing that "with paradigm changes in the humanities and social sciences . . . attention has focused on culture, and cultures, as fluid and shifting, rather than as bounded wholes moored to localities that define them".

These evolving perspectives bring to light more than the multiple shifting influences on culture and their myriad manifestations. They also prompt attention to regional variations in cultural epistemologies and ontologies, and at their logical extensions, questions about the very notions of culture that they attempt to elucidate – particularly as cultural realities are expressed more as systems of becoming than as systems of preservation.

Indentureship and Cultural Transplanting

Writing with reference to the diasporic settlement of Indian indentured labourers in Fiji and South Africa, Reddy (2015, 2) suggested that "Indians are othered by host nation states because of their perceived transnational identities (in this case, their Indianness), and they inadvertently distance themselves from locals by emphasizing their ties to India, which over generations is weak".

In Trinidad of the nineteenth and early twentieth centuries, the transplanted Indians and even their children born in the adopted home were physically and culturally foreign. Sookdeo (1988, 30) noted that the new arrivals were bound by contracts that followed the old master-slave ordinances and were kept in what were known as "nigger yards" and that the freed slaves looked down on them, resulting in such perceptions as "'the starving coolie', 'job stealers'; and other negative stereotypes".

Balliger (2006, 281–82) suggested that global influences including the distant impacts of policy changes within the British Empire and local politics in India encouraged the system of indentureship and also entrenched attitudes to the new arrivals that persist into modern society and culture. As poverty and famine pressed upon parts of India, the British used migration as a means of capitalizing on the desires of the lower classes for economic advancement and to help ease tensions in troubled areas. The Indians who responded to the British inducements to migrate to the Caribbean found themselves in a foreign environment where they "undercut wages and empowerment among freed African slaves, creating resentment and hostility along 'racial' lines that continues into the present" (ibid., 282).

Manuel (2000a, 4) similarly noted that "Indians were regarded with contempt by society at large" such that "whites saw them as coolie heathens who drank excessively, abused their wives, dressed like savages, and clung to their backward ways and pagan faiths" while "most black West Indians, long since Christianized and largely alienated from their ancestral African roots, shared these colonial prejudices". Similarly, Kelvin Singh (1996, 231) has argued that the white plantation owners and colonial rulers used race or ethnicity as the "fundamental organizing principle" of the colony such that "consciousness of race permeated the whole society" and conditioned both the image of the Indian arrivals as perceived by others and by themselves such that "their initial entry into the society as indentured labourers, despised even by the recently emancipated slaves and their immediate descendants, coupled with the Indians' equally despised "heathen" or "pagan" religious beliefs and practices in a period of fervent Christian evangelization, inevitably made all Indians acutely aware of their degraded social status".

For Jain (2004, 9) the attitudes of the new arrivals were partly the result of their situation in that the indentured labourers, "believing their situation in Trinidad to be only temporary, displayed little enthusiasm to integrate with the wider community". However, Jain (ibid.) also noted that "the conditions imposed by the colonial government to confine the indentured labourers to the plantation environment of the sugar belt, together with the black population's hostile attitude toward them" left them with "little alternative but to remain exclusive".

The first Indo-Trinidadian prime minister of Trinidad and Tobago, Basdeo Panday, has been quoted as acknowledging the early Indian immigrants' propensity to remain apart from the creole colonial society, saying that "Indians came here with a return psychology . . . Right up to 1940–45, there were Indians here who had the idea that they would go home to die. And it was because of this mentality that they did not participate" (cited in Ela Dutt, "Tension over Indian Aspirations", *India Abroad*, 12 November 1993, 24).

Brereton (1979, 2) characterized this lack of integration in terms of not just the choices of the indentured arrivals and subsequent settlers, but also as a function of the perceptions of the earlier creolized elements of the emerging society, writing that for many decades after their arrival, "Indians remained marginal to Creole society", from which "they were viewed as a group of migrant labourers, birds of passage, who would not remain to form a permanent part of the population" even after it became clear that large numbers of them were indeed to stay and settle in Trinidad.

Munasinghe (2001, 89) argued that notions of exclusivity and apartness would continue despite increasing social mobility and involvement after the end of the nineteenth century. The case for this continuing apartness may be somewhat overstated, but the evolution of the Indo-Trinidadian sphere of activities into the broader society has continued. Jain (2004, 14) argued that, outside of the internal dynamics of Trinidad's community, global forces also fostered the movement of Indo-Trinidadians away from agriculture and into a broader range of social and economic activities, observing that "the socio-economic mobility of the East Indian community, along with that of the other ethnic sections of Trinidad, was accelerated during periods of the so-called American 'occupation' of these islands during the Second World War and, more recently (1974-83), with the oil-boom".

Analyses such as those of Munasinghe (2001) and Jain (2004) depend on relatively orthodox notions of culture and identity, inextricably linked with notions of ethnicity that have traditionally dominated almost all investigations of the Indo-Trinidadian presence in Trinidad. However, modern globalized communication flows and influences suggest that identity choices now become more available and relevant in the determination of social and cultural place. The post-cultural construction of identity may now depend as much on the cultural choices available to an individual as on that person's immersing native culture, constituting what Manuel (2000a, 203) has characterized as "projects of identity construction", including the influence of Saudi Islamic thought in Trinidad and Tobago in which Arab-rooted Islamic ideas and culture are supplanting the Indo-Muslim traditions such as the *maulud* and *qasida* musical forms.

Global Identity Markets and Cultural Invention

For several years scholars like Anderson (1983) and Sollors (1987) have advanced the notion that culture, identity and nationalisms are not, in fact, fixed, traditional, inherent and inherited, but often rather recent, pragmatic and convenient conceptual inventions (or reinventions) that suit contemporary needs, spread by common understandings (whether accurate or not) of cultural myths and shared symbolic meanings. These ideas about imagined communities have found resurgence in studies of nationalism and diaspora in the context of digital global communication technologies (see, for example, Candan and Hunger 2008; S.N. Mohammed 2012b). These pre-Internet-era ideas have flourished as access to global communication technologies has enhanced the ability of diasporic groups and other nation-constructs to share and construct meanings through media content and personal expressions in an unprecedented manner and thus to engage in the processes of imagination and construction of identities and nationalities.

The increasingly diverse interplay of influences in the globalized communication and cultural environments has led many modern scholars (see, for example, Bhabha 1994; Burke 2009; Kraidy 2005) to focus on the concept of "hybridity". Trinidad and Tobago's multiple cultural legacies include remnants of several traditions, including West African, Indian, Chinese and European influences. These traditions have, inevitably, intermingled.

Mathews (2000, 1) argued that culture can continue to be a meaningful concept if modern definitions integrate the notion of culture as "the way of life of a people" with the notion of culture as "the information and identities available from the global cultural supermarket", while Bayart (2005, x) suggested that most of our modern societal issues involve what he terms "the problems of the illusion of identity", wherein, somewhat paradoxically, even "the general opening up of societies – 'globalization' – is accompanied by the exacerbation of particular identities, whether religious, national, or ethnic".

Hall (1997, 2010) suggested that while modern identities may well involve conscious and transient choices, they are not completely free of the pressures of established stereotypes and representations. Hall described such claims in terms of a continuing struggle between identity assertions and the hegemonic forces that serve to force established and stereotypical representations onto particular groups or individuals.

Hybridity, Authenticity and the Anterior Pure

Whereas cultural phenomena such as Trinidad's chutney soca (a fusion of musical styles from Indian and African traditions) are easily classified as hybrids, the concept of hybridity has grown beyond simply describing instances of fusion or mixture. For Kraidy (2005, 75), hybridity has moved beyond a description of particular instances of cultural commingling towards the ascension of the broader concept that "all cultures are hybrid", which he argued "is clearly ascendant, and even nearly consensual, in intellectual and public discourse".

Technologies such as satellite television and the Internet are often central to investigations of hybridity – but even where these technologies are not central concerns, the influence of global communications systems is never far from the thoughts of hybridity theorists. Kraidy's work on Arab reality television, for example, not only pivots on the popularity of satellite television in that region, but also implicitly depends on the international communication and entertainment currents that have developed, popularized and sustained the genre of reality television.

Scholars have applied (and critiqued) hybridity in a variety of modern cultural phenomena, including Bollywood weddings (Kapur 2009), reality television in Saudi Arabia (Kraidy and Khalil 2009), Asian game shows (Hetsroni 2005), Chinese martial-arts films (Wang et al. 2008) and Arab satellite television advertising (S.N. Mohammed 2012a). Burke (2009, 2) argued that regardless of one's view of hybridity, it is impossible to miss. Notably, McFarlane-Alvarez (2007, 53) found hybridity in as minor a detail as commercial breaks on television in Trinidad, which featured not only a variety of international products and influences, but also a variety of Trinidadian people and cultures, concluding that "television advertising in Trinidad and Tobago is a space of hybridity and liminality, through which the Caribbean culture of *mestizaje* is expressed".

The representational hybrids in McFarlane-Alvarez's (ibid., 52) reckoning even included the people of Trinidad and Tobago, since "television advertising in Trinidad and Tobago uses hybridity of representation to interpellate the viewing audience" such that "in the Trinidad and Tobago context, talent of mixed African and Indian descent (referred to colloquially as *douglas*) are [*sic*] used to represent both African and Indian populations, or a 'middle brown' or 'red' mixed actor might be used to represent middle Trinbagonianness".

Tanikella (2003) is among those who have argued that any analysis of hybridity also demands an interrogation of the power relationships between the groups and influences involved. From this perspective, cultural hybrids are not just simple manifestations of cultural mixing, but are also embodiments of the global, historical and political forces that determine whether such mixing occurs, the terms and outcomes of the mixing and the acceptability of the mixed form (Gilroy 1993; Puri 2004; Ramnarine 2011; Tanikella 2003).

As a further step, it becomes clear that interrogation of the power relationships involved in hybridity also demands that the historical context of the particular mixings be examined. Both Ramnarine (2011) and Nederveen Pieterse (2001) hinted at the importance of historical context in analysis of modern hybrids. Nederveen Pieterse (ibid., 221) emphasized the importance of "old hybridity", or the pre-existing social and cultural conditions that govern the development of new hybridities. Ramnarine (2011, 143) pointed to several examples of "new" diaspora communities such as Indo-Trinidadians in North America whose hybrid identities can be traced to earlier historical cross-cultural experiences (indentured servitude under British colonial rule brought their ancestors from India to Trinidad) in what she terms the "old diaspora" experience. These observations suggest that scholarship on globalization and culture is coming to terms with the idea that cultural mixings – whether considered hybrid, creole or multicultural – are complex interactions, conditioned by both power and history. Perhaps more importantly, emerging approaches to global cultural change are being forced to engage with persistent issues of hegemony that condition the much-touted hybridities of the global information society and the extent to which the globalizations of today may well reflect many deeply embedded globalities of the past.

Despite its popularity, the concept of hybridity presents several difficulties, not the least of which is its lack of resonance with the groups and cultures it claims to analyse. Straubhaar (2008), for example, pointed out that people are not likely to identify themselves as cultural hybrids, even if they admit to having multilayered or diverse identity claims. In part, the hybrid subjects of the academic gaze may themselves find the notion of a hybrid less than desirable, particularly since identity claims, group membership and cohesion have been (and continue to be) important social factors.

The hybridity perspective also fails to account adequately for the diversity of historically meaningful identity cues (including global forces such as colonialism) that have constituted so-called traditional identities. Thus the presumably pure or authentic cultural identities that are conceived of as being hybridized may themselves be hybrids that emerged from the interplay of historical cultural intermingling. Afro-Caribbean cultures, for example, have arisen out of the European slave trade, the experiences of slavery, colonialism and modern global forces. To characterize the cultural transformations observed in their societies today as a result of US media influences as hybridity misses the already hybridized nature of their cultures.

Research has also demonstrated that hybridity may not necessarily preclude traditional identities. Several investigations have concluded that groups may demonstrate outward shows of hybrid influences while retaining traditional core values and identities. Mohammed and Queen (2011), for example, reported first-person accounts of young people in Kuwait who felt that their core beliefs and values could not be influenced by Western culture even if they adopted Western dress and other outward symbols to express themselves. Scrase (2002), in an investigation of Bengali television audiences, argued that members of the Bengali middle class demonstrated a tension between their core traditional cultural identities and the more contemporary identities they adopted (or appeared to espouse) from exposure to foreign television.

Puri (2004) went a step further in the interrogation of the hybridity concept, arguing that the term hybridity is itself value-laden with particular paternalistic ideas that create bias in analyses falling under its scope. For Burke (2009), the proponents of hybridity are, notably, themselves hybrids. This observation would suggest either that they are more sensitive to hybridity or that they are simply more likely to see global cultural developments in terms of hybridity. Burke (ibid., 13) wrote that "examples of cultural hybridity are to be found everywhere, not only all over the globe but in most domains of culture – syncretic religions, eclectic philosophies, mixed languages and cuisines, and hybrid styles in architecture, literature or music".

The entrenchment of hybridity as a given (what Kraidy [2005] has characterized as a "master trope") in discussions of global cultures raises several questions about its underlying assumptions. Hutnyk (2005, 81), for example, asked the key question: "To what degree does the assertion of hybridity rely on the positing of an anterior 'pure' that precedes mixture?" Much of the conceptualization of hybridity and of culture more generally involves implicit assumptions about something called authenticity. Lindholm (2008, 2) noted that cultural authenticity is difficult to define, but offers the assertion that "authentic objects, persons, and collectives are original, real, and pure; they are what they purport to be, their roots are known and verified, their essence and appearance are one".

Richard Handler (1986, 2) characterized authenticity as a particularly modern quest for elusive values that may exist more in the imagination than in anyone's lived experiences, "a cultural construct of the modern Western world". Authenticity is predicated on the presumed separate and pure origins of various cultures – a condition which may exist in the rare case of isolated communities, but which in the vast majority of cultures now known does not hold true. Despite this, notions of authenticity form a major (if sometimes implicit) part of modern cultural analysis. A cross-database search indicated that "authenticity" and "culture" appear together in approximately sixty thousand scholarly journal articles from 1970 to the present. This association bears further investigation, particularly since authenticity, like culture, remains ill-defined and variously applied.

A focus on authenticity in certain contexts where differing cultures face each other – notably tourism – demonstrates concern about how one culture may represent itself to another (or to several others). For MacCannell (1973, 589–90), a search for "authenticity of experience" is manifest everywhere in modern society and "the concern of moderns for the shallowness of their lives and inauthenticity of their experiences parallels concerns for the sacred in primitive society".

Bruner (2005, 4–5) argued for avoiding authenticity altogether, writing that "authenticity is a red herring" belonging to "anthropology's discarded discourse, presenting cultures as functionally integrated homogeneous entities outside of time space and history". Note here that, for Bruner, authenticity invokes a specious historical fixedness.

In the particular case of Trinidad and Tobago, both the notion of authenticity and the notion of the anterior pure are relevant and thorny issues. Green (2007), for example, described the efforts of government agencies in Trinidad and Tobago to market the country's annual pre-Lenten Carnival celebrations to tourists abroad, in which efforts they sought to distinguish the authenticity of the Trinidad Carnival event as the "real" or "true" Carnival experience. Green (ibid., 213) noted that the common strategy of describing Trinidad as the "Mecca" of all carnivals represents an attempt to "claim ultimate authenticity". Green (ibid., 220) also went on to argue that these claims involve economic repercussions, since "commerce and culture are linked through a moral project of nation building that relies on an international politics of recognition grounded in the 'possession' of 'authentic' cultural forms".

At the same time, the internal competition for (however specious) authenticity continues as well, with noted Caribbean personalities such as Rex Nettleford (1998, 193) having suggested that Carnival and other typically Afro-Caribbean observances can lay greater claim to "authenticity" than Indo-Caribbean observances:

By the time the latecomer East Indians, Chinese and Lebanese from the Levantine Coast entered the region, the rules of the game had been made; not even the overwhelming majority East Indian population of Guyana and the sizable minority of the same group in Trinidad has jerked these countries out of their historical Euro-African or Afro-Creole realities. Carnival and *Jonkonnu* are unashamedly Afro-Creole or Euro-African expressions, claiming a particular authenticity over *Divali* (Festival of Lights) and *Hosay* as genuine ancestral Caribbean expressions.

In the Indo-Trinidadian community, these issues of authenticity and the anterior pure are particularly difficult ones. Considerable debate has existed for many decades in that community over what constitutes authenticity and what is pure. Yet, at the same time, there exists a pervasive awareness of the hybridity or adaptedness of those things perceived as constituting local "Indian" culture.

We have seen before that among the most important markers of Indianness for the majority of Indo-Trinidadians is the practice of Hinduism, perceived as a single orthodoxy, brought whole in its original form from India. The term "Hinduism" to describe Indian religious practices spread and varied over the subcontinent is a clear British invention dating back to 1829 (Lorenzen 1999), though the practices and beliefs it describes have indeed existed in regional variations for millennia. Scholarly debate rages on over whether Hinduism was an invention of the colonial gaze on the basis of the fact that the practices existed prior to the coining of its name.

Lorenzen (ibid., 631) countered that, based on the evidence, "a Hindu religion theologically and devotionally grounded in texts such as the *Bhagavad-gita*, the *Puranas*, and philosophical commentaries on the six *darsanas* gradually acquired a much sharper self-conscious identity through the rivalry between Muslims and Hindus in the period between 1200 and 1500, and was firmly established long before 1800".

Whatever the origins and state of Indian Hinduism, the evolution of the particular set of religious and cultural ideas under this label evolved somewhat independently in Trinidad. Edmonds and Gonzalez (2010, 179) wrote that "Hinduism in Trinidad became quite distinct from Indian Hinduism, incorporating new myths, rituals, and festivals", arguing, as well, that "the regional plurality of Indians and their distinctive styles of worship eventually came together in the Caribbean and formed a more generalized, less regionally specific Hinduism". These tensions of tradition and innovation are most evident and most commonly discussed in the observances that draw on or claim association with Indian heritage, including communal events such as Divali and Phagwa (Holi) and family events such as weddings and prayer ceremonies. Writing in the 1970s, Jha (1976) noted that, even in formal rituals associated with their Indian heritage, Indo-Trinidadians faced the morphology of their practices and the attrition of their connections with (an also-changing and at times nebulous) Indian Hindu orthodoxy.

While the struggle for authenticity and the quest for the anterior pure dominated many cultural events, the countervailing forces of hybridity and syncretism were also broadly evident. Korom (2003, 98) has noted that the Muharram ritual called Hosay in Trinidad (itself a deeply hybridized observation emerging from Shi'a Islam in Persia arriving in the Caribbean via centuries of morphing and travelling through South Asia) provides an example of this mixing, particularly as a visible show of cultural presence.

Richman (2010, 84) captured something of the complicated web of influences, including factors present, past, near and distant, that influenced the negotiation of identity post-indentureship to include such mundane changes as cheaper travel, which allowed for the exchange of cultural ambassadors from India to Trinidad and Tobago and tourists from Trinidad and Tobago to India who, while in search of their roots, "encountered not the idealized Hinduism of their grandparents' stories, but the complex, fluid, and often wildly variant Hindu practices of 20th-century northern India".

To a lesser extent, educational exchanges also provided some connections between Trinidad and India. There were Indo-Trinidadian academics who completed their graduate studies at Indian universities, sometimes with the involvement of government or intergovernment agencies and at times with the support of the Indian High Commission. Notable academics and religious leaders in the Indo-Trinidadian community such as University of the West Indies lecturer Brinsley Samaroo and head of the Hindu Prachar Kendra (Missionary Institute) Ravindranath Maharaj (Ravi Ji) have completed at least part of their education in India. Modern connections with India as imagined homeland or point of cultural origin are both more plentiful and more diverse, with modern media and business connections. They include broader media contact than Bollywood films – today including soap operas and news from India – as well as business ventures such as so-called Indian Expo fairs.

Conclusion

The theoretical positions outlined here present an interesting set of possibilities against which Indian-format radio in Trinidad and Tobago may be examined. Questions about the manifestations of hybridity as well as explication of the numerous global and historical forces impinging on this media practice are open ones requiring analysis. Similarly, these theoretical perspectives beg the question of how these media practices, including social media components of the stations, are involved in identity processes. How, for example, do listeners engage with content that they do not have the linguistic ability to understand? How, also, does a media sector that caters to one part of the population figure into the politics of national identity, and to what extent do the stations featuring Indian content also lay claim to being fully national in their scope? These are questions to which we will return after examining the genesis and evolution of mass and electronic media and Indian-format radio in the region and in Trinidad and Tobago.

CHAPTER 3

Early Radio in the Caribbean

The Indian-format radio sector in Trinidad and Tobago emerged out of a long and diverse historical tradition of mass media in the Caribbean region. When the first of these stations emerged, it did so into an electronic-media environment that could trace its origins to within a few years of the start of broadcast radio. Though Trinidad would not have its own radio station until 1947, radio was an emerging social force from well before that time. Wired radio-relay systems were part of the local listening environment as the radio technologies evolved, but wireless enthusiasts refused to be limited to the feed on the wire. Like their counterparts in many parts of the world, radio listeners in Trinidad tuned in to distant signals from wherever they could receive them. Many Indo-Trinidadian listeners at the time could still understand something of their ancestral Indian languages and discovered that they could receive content in that language, and consonant with their cultural expectation, from distant sources like All-India Radio and from some less distant sources such as Guyana, where local broadcasters were already into the business (and it was a business) of Indian content on the radio.

Caribbean Radio in the Global Context

Early radio involved experimentation that pushed the boundaries of the possible. Point-to-point radio communication, at first using Morse code (and only later actual audio), was among the first uses to which the new technology was applied. The idea of broadcast radio intended for a wide anonymous audience emerged out of official systems that transmitted information to the public, such as the US Naval Observatory, which, starting in 1913, broadcast time signals over the air to anyone with a device that could receive them. The development of radio also benefited from the informal work of radio amateurs who dabbled in transmitting sound to other enthusiasts and to small communities of enthusiasts around them. These amateurs continued to experiment with radio even as governments began to issue broadcast licences to formal radio operations such as KDKA in Pittsburgh in 1920 and the British Broadcasting Corporation (BBC) radio service in 1922.

Fejes (1986) has argued that amateur radio enthusiasts in the early 1920s surprised and challenged science when they proved that radio waves could be transmitted and received over such long distances that global reach became a possibility, developing radio beyond its point-to-point private-message function into a broadcast entertainment and information mass medium that was itself a powerful tool of empire and global reach, of interest to colonial powers and emerging superpowers alike.

For British broadcasting, early impetus came from military demands during and just after World War I. As Anduaga (2009) has described it, the British government established numerous research institutions in order to advance scientific knowledge and to improve security. Among these security concerns were practical matters such as maritime and aviation navigation, in which radio technology played important parts. The emerging global scale of wireless broadcasting also became an important consideration as research institutes in Britain paid special attention to mapping as widely as possible the variables affecting how far and how well signals were being received on a global scale.

The distance-bridging potentials of wireless commercial broadcasting presented possibilities of massive scale to the British Empire and other global forces of the time. Hitherto-limited information flows achieved through such technologies as undersea cables were being overtaken with powerful instantaneous multipoint broadcasting that could cover great distances and reach multitudes. As with many historical and technological changes, this push to use radio to unite the empire was not purely the result of radio technology, or even the specific emergence of shortwave technology with its improved reach and coverage, but also, in part, the extension of previous such efforts.

Among the earlier efforts at such connectivity, various agencies and government departments worked towards the establishment of what was to be known as the *Imperial Chain,* which sought to connect the major components of the British Empire with the technologies that existed during and just after the First World War. This Imperial Chain was approved at the Imperial Conference in Melbourne in 1909 ("Wireless Telegraphy: Marconi Contract Accepted", *Times of India,* 9 March 1912, 9) and received considerable attention as wireless telegraphy technologies evolved. Colonial authorities saw tapping this potential to unify the empire as an important move both for political and for military purposes.

Much of the inability to advance these plans and make the systems feasible had to do with the power of the British Post Office, which exercised bureaucratic domination of communication systems. In 1913, the British postmaster general (then Herbert Louis Samuel) appointed an advisory committee that was "requested to report, within three months, on the merits of the existing systems of long distance Wireless telegraphy, and in particular as to their capacity to fulfil the requirements of the Post Office for the proposed Imperial chain of wireless telegraphic stations", noting: "These requirements involve continuous communication between station and station by day and night over land and water and over distances of between 2,000 and 2,500 miles" (Advisory Committee on Wireless Telegraphy 1913, 3).

Far from being confident of the chances of establishing the Imperial Chain at the time, the committee warned, instead, of the wide range of proprietary, mutually incompatible competing technologies currently in use, with several firms (including foreign companies) being involved. Committee members argued that with this mish-mash of proprietary technologies, the various companies struggled to not step on each other's toes in avoiding the use of technologies that might be protected by competitors' patents. They concluded that what existed in the Imperial Chain did not properly constitute a "System of Wireless Telegraphy", writing (ibid., 4):

> The main differences in the apparatus and devices at present in use centre on the high-frequency generator; but there are other apparatus or devices protected by patents, such as special aerials, different types of automatic high speed transmitters and of receivers and recorders. Any company or firm which makes use of apparatus or devices which others are precluded from using because of the existence of some patent may claim to have its own system of wireless telegraphy, but the term "system" is misleading.

Despite these doubts, the committee was convinced that at least some of the technologies that its members examined would soon enable long-distance wireless telegraphy. After many years of bureaucratic and technological wrangling, the Imperial Chain would eventually be set up in 1927 (Shoup 1928), by which time broadcast radio had already partly eclipsed its usefulness as a source of public information. Shortwave transmissions, when developed, formed part of the Imperial Chain, which would later evolve into the British telecommunications giant Cable and Wireless.

Thus the British Empire had, for several decades before the introduction of long-distance shortwave wireless radio broadcasting, sought to establish an extensive and immediate broadcasting and communication presence in the colonies. Notwithstanding the existence of the Imperial Chain in its particular form – that is to say, primarily as a point-to-point medium for private communications – shortwave radio presented new and exciting possibilities. As Anduaga (2009, 67–69) has noted, the political and the commercial potentials of wireless radio broadcasting were intertwined, capturing the imagination of the colonial authorities and the colonial populations as well:

> Underlying the booming of short wave, however, were commercial imperatives. And underlying these was, among other aims, an ideal that was always intimately associated with the wireless development: Empire's unity. . . . The late 1920s witnessed a marked idealization of short wave as a vehicle of unification over large areas of the British Empire. The journals and magazines of the period show an increasing tendency towards a kind of devotional acceptance, an appeal to the fraternity through waves. On the one hand, the notion that, thanks to short wave, something happening in Britain could be transmitted – the mere possibility of hearing voice! – and experienced in the Dominions, and vice versa, was completely new and exciting.

The commercial imperatives involved in radio were sometimes more parochial than the broader concerns of Imperial welfare. Among those exploring the connections between global power and commercial interests in the spread of radio, Bowden, Clayton and Pereira (2012, 931) have described the interplay of commercial and political imperatives in the diffusion of wireless radio technologies in the early 1900s, noting that initial efforts were primarily aimed at supporting the emerging British wireless radio manufacturing industry:

> Radio was a part of the electronics sector of British industry – a sector that was one of the fastest growing industries in the field of manufacturing, and also one in

which exports were increasingly viewed as essential to the health of the economy. Government policy, designed to assist the growth of the industry, particularly in relation to export markets, was part of the overall concern to promote the development of the electronics industry. Government policy was based on the assumption that consumer demand for wireless radio receiving sets in the colonies would rise, and that this would encourage British manufacturers to mass-produce for such markets.

In addition to the promotion of wireless radio within the colonies as a function of the desire to support the British electronics sector, as might be expected, there is evidence that radio was also used to serve the broader economic interests of the British Empire. Radio was, for example, one of the tools that the Empire Marketing Board used within the United Kingdom to promote trade with and within the colonies up to the early 1930s. Stephen (1933, 255) described the board's efforts in the following terms: "Every conceivable modern publicity device has been seized on by the Board. Each month last year 200 Canadian films were shown to audiences with an average monthly total of 46,000. Housewives were attacked over the radio in a series of talks describing week by week Empire products that were in season."

Bowden, Clayton and Pereira (2012) noted that these economic imperatives were only part of the equation for radio in the British Empire, as this medium was also seen as part of a civilizing mission for the colonies and one which sought to provide information security for the empire. In the emerging global information environment, populations of the far-flung corners of the British Empire were gradually encountering a wider range of information sources that might be traditional printed materials or word of mouth, but could also be radio broadcasts from competing powers. Bowden, Clayton and Pereira (ibid., 932) noted that the British Empire's radio strategy started with bridging electronic divides "to provide colonial peoples with access to radio broadcast technologies" but continued with "the specific remit of controlling the information the colonial countries received" such that "British policy makers had two strategic aims. The first socio-strategic aim was to ensure that colonial audiences received 'reliable' and 'true' (that is, their definition of reliable and true) sources of factual news and information. . . . The second geo-strategic aim was to ensure that this information communication technology transferred and extended 'British values' amongst the colonial populace." Various global forces thus came to bear on the development of radio as a global medium, including the changing international environment leading up to and including World War II. This would eventually mean that the strong emphasis on purely British imperial concerns would also expand to accommodate more global concerns. In particular, the impact of World War II saw not only a shift in British radio strategies but also the involvement of US radio broadcasting to counter common threats. As Fejes (1986, 63) recounted, the American government, initially not eager to use shortwave radio as a means of furthering its foreign-policy objectives, eventually warmed to the idea of this technology, particularly with the prospect of international competition for influence in neighbouring regions:

> Initially, the development of a program of shortwave broadcasting to Latin America was seen as forming closer ties between the United States and Latin America. Such broadcasting would play a role in the construction of an inter-American system of political and economic relations organized around the goals and interests of the United States. However, with Germany beginning to threaten the dominant position of the United States in the hemisphere in the latter part of the decade, the administration began to regard as urgent the development of shortwave as a method of countering German threats.

However, despite American involvement in radio in several instances, it was British radio that dominated the Caribbean region from the start. Dunn (2004a, 76) wrote, for example, that by the time radio began to take root in the Caribbean, "London was already exercising uncontested political control over the region and its nascent electronic media infrastructure" and that "early public radio communications facilities" in the region "were meant to maintain the morale of the population and provide government information during the Second World War". This use of radio as a tool of empire was, perhaps, not surprising, given that the anglophone islands were, at the time, British colonies.

In most cases, some combination of local (often experimental) infrastructure (such as private amateur stations) and British capital gave rise to formal broadcasting operations which at the time often included wireless broadcasts and wired transmissions on what was known as the Rediffusion system. These various arrangements led to the establishment of several broadcast operations throughout the region, including Barbados in 1934 and Trinidad in 1947. *Variety* ("Trinidad's Commercial Station Preems in August", 25 June 1947, 26) reported the launch of Radio Trinidad in a report datelined 12 June as follows:

Trinidad and Tobago governments have given franchise to the Trinidad Broadcasting Co. to operate a commercial station, and new station will get under way shortly at Caroni, nine miles from Port of Spain. Station, known as Radio Trinidad, will operate on both medium and shortwave bands, planning to be heard also in adjacent Caribbean colonies and parts of South America. . . . Station hopes to open by mid-August. There'll be regular daily transmissions 7 to 9 a.m., noon to 2p.m. and 4 to 11 p.m., with extension on Sundays between 9 a.m. and noon. Test transmissions will start about July 1. . . . Acting general manager will be Frank Lamping. Latter, senior exec of International Broadcasting Co. of London, expects to go to South Africa in late August, and will turn over Trinidad post to W.A. MacLurg, formerly with BBC.

Clearly evident in this report is the colonial scope of the enterprise, which envisioned regional broadcasting to nearby colonies (such as Guyana) and involved British radio officials who exercised managerial and political power over the broadcasting operations.

While it is widely recorded that the BBC initiated experimental overseas transmissions in 1927 (Fejes 1986, 72), to date the history of radio in the region as somehow starting with formal broadcast operations sanctioned by the British authorities would be a mistake. Efforts at broadcasting in the region originated with hobbyists and experiments and only later became a priority for the empire (based partly on the geopolitical realities of the time, including broadcasts by Nazi Germany's propaganda ministry starting in the early 1930s).

The Guyanese experience with broadcasting is of particular relevance to the present discussion for its early broadcasting experimentation and because of its direct and indirect connections with Trinidad and Tobago radio generally and Trinidad Indian radio in particular. Sanders (1978) wrote that within four years of the start of BBC (then the British Broadcasting Company) transmissions in 1922, Georgetown telephone subscribers were able to listen (for a small fee) to two hours of programming per week, received via shortwave from England. Sanders (ibid., 20) suggested that "the consumers of the radio transmissions were expatriates who would have been interested in the BBC transmissions and indeed who were afforded, and could afford, a telephone or receiving set".

Veteran Guyanese broadcaster James Sydney (n.d.) noted that the telephone relay system was abandoned after an experimental shortwave wireless facility was put into service, and that from 1927 experimental shortwave broadcasts were introduced for two hours a week. According to Berg (2013, 64), in 1928, station VRY, Georgetown, British Guiana, could be heard two nights a week on 6850 kilocycles. Sydney (n.d.) mentioned that these broadcasts continued until 1931, when funding problems brought them to a halt, though Berg (2013, 81), who referred to VRY as "the widely-heard station in Georgetown, British Guiana", placed the station's closing in 1930.

It was the appeal of cricket that would revive broadcasting in Guyana. As Sanders (1978, 21) recounted, the revival of radio broadcasting in Guyana was prompted by the visit of the Marylebone Cricket Club from England, since "broadcasting was revived in February 1935, by a group of enthusiasts in time for broadcasting a ball-by-ball commentary of the cricket matches played between the visiting MCC [Marylebone Cricket Club] team and colony teams".

Berg (2013, 127) noted the return of radio in British Guiana in 1935, but with no mention of the role of cricket, writing that "British Guiana returned to the air in the form of VP3MR, the British Guiana Broadcasting Company, on 7080 kc. The 150-watt station operated for an hour or two most days around 1700-1900 and Sundays at 0745-1015". Berg (ibid., 143) also noted that in 1936 "a second station came on the air from British Guiana. It was VP3BG, 'the Voice of Georgetown', on 7220 kc. It was owned by the Crystal Broadcasting Company, and could also be heard in amateur operation on 20 meters. This station had facilities at 1, Wellington St., Georgetown." Sanders (1978, 21) characterized the evolution of the two early British Guiana stations stemming from the coverage of the Marylebone Cricket Club matches, writing that "two stations, VP3MR and VP3BG, were operated independently from this time on a commercial basis with sponsored programmes until 1938 when they were amalgamated on the formation of the BG [British Guiana] United Broadcasting Company Ltd, which was financed by local firms and individuals. . . . In 1949 a medium-wave transmitter was brought into service in addition to the shortwave transmitter." The *New York Times* ("Radio's Short Waves", 17 January 1937, 164), mentioned these Guyanese stations among those radio signals that were being received in the continental United States on the shortwave band in those days: "Station VP3MR at Georgetown, British Guiana, is heard clearly these days on its new frequency of 5.98 megacycles. With the exception of VP3BG on 6.135 megacycles, VP3MR is the only English-speaking short-wave broadcaster on the continent of South America."

But even before any broadcasts originated from British Guiana, locals there took an interest in radio and the signals they could receive from distant lands.

While local newspapers were still generally unaware of the several private experiments in radio reception in places such as Guyana and Trinidad, foreign newspapers were intrigued. The *Chicago Daily Tribune* took note of this fact in a brief article that reads in part:

> There's a man in the jungle, 200 miles from the nearest habitation in British Guiana, who, when he asks his supervisors off in Georgetown for a short vacation for a brief brush with civilization, is always told: "You get more amusement out there than we do here in town." The reason is this: H. Johnstone Smith, Inspector of police at Morewanna, British Guiana, is constantly entertained by concerts broadcast from KDKA at Pittsburgh 3,000 miles away. ("Hears KDKA on One Tube in British Guiana Jungle", 9 November 1924, F10)

The paper's very reporting of this reception and its emphasis on the distance involved in Smith's enjoyment of these concerts suggest that this was a novel observance. The notion of a civilizing mission is also evident in the suggestion that radio provided a brief contact with civilization by providing concert music to someone in the jungle.

By 1938, radio was a regional phenomenon in the Caribbean but also an international phenomenon that was aware of the global forces at play in its production and reception. A particular testament to this global view emerges from an article in the *Times of India* (29 November 1938, 19) regarding the impact of All-India Radio that made specific reference to the Caribbean. The article read, in part:

> Evidence is growing indicating the manner in which the broadcasts of the All-India Radio are being appreciated by Indian settlers in various parts of the world. In East Africa, South Africa, Ceylon and Malaya there are numerous Indian and other listeners who tune in to Indian stations with good results. But this is not necessarily due to nearness of these countries to India. Even such distant lands like British Guiana and Trinidad are in a position to contact Indian stations with considerable results. ... As such the letter sent by an Indian resident of British Guiana to the Director of the All-India Radio suggesting certain ways and means for the more distinct reception of Indian stations in the West Indies and for the enlarged range of broadcasts from India for the special benefit of Indians overseas, is of particular interest.

The letter recounted in the article goes on to indicate that, despite some difficulties with reception, listeners in British Guiana looked forward to listening to the broadcasts from India which were received from 9:45 local time, noting (ibid.), "Your programme is then appreciated and creates a happy feeling in the minds of overseas Indians."

By that time, Guyanese radio listeners had access to a combination of local and global programming, while those in Trinidad were listening to radio from international and regional sources (including Guyana). However, despite the fact that there were relatively few radio sets spread among local families (or perhaps because of this fact), radio listening turned out to be a noisy affair that was popular (and loud) enough to draw the ire of the authorities. In both territories, authorities imposed a listening curfew described in *Variety* magazine:

> Receiving sets tuned in after 11 p.m. in Trinidad and British Guiana, South America, get the listener, if nabbed, the alternative of forking up to a $25 fine or cooling his heels in the clink. It's against local anti-noise ordinances. Curfew is strictly enforced by the gendarmes and violators are dealt with immediately. Rule covers morning listening as well, 6 a.m. being the earliest for pickups . . . Exceptions are allowed only on certain occasions and then only by official sanction of the Wireless Authority which runs okay notices on the front pages of local dailies. Then it's only for important sports events, etc. shortwaved from the U.S. or England. ("Tough Curfew in British Guiana", 1 June 1938, 38)

Sydney (n.d.) wrote the following about the regional scope of the shortwave broadcasts from ZFY in Georgetown (later to become Radio Demerara), which continued to have an impact even after the start of radio in Trinidad: "Interestingly, ZFY had a significant Trinidad audience. For many Trinidadians, it was the main or only source of religious broadcasts and of Indian musical entertainment. Even after September 1947, when Radio Trinidad was inaugurated, ZFY retained a sizeable Trinidadian listenership."

The geopolitics of the day also fuelled with some intensity the popularity of broadcasts aimed not only at US Armed Forces but also at audiences for wartime information and propaganda. Well before the launch of Radio Trinidad, New York's *New Amsterdam News* of 4 May 1940 reported that Port of Spain, Trinidad, had experienced "tumult" in the past week due to the confiscation of radio sets for non-payment of licence fees (which amounted to $2.40 for the year) and noted the connection between these listeners and the events of World War II: "With the superabundance of propaganda and the fact that American commentators seem to know more about the war in Europe, than even the English, the West Indian's only authoritative source of information is what he gets from the American short-wave stations" (Phyllis H. Matthews, "In Caribbean",

New York Amsterdam News, City edition, 4 May 1940, 8). Additionally, the *New York Times* later that same year noted the operation of several stations in the West Indies and South America, including those at St Kitts, British Honduras, Nassau, Bermuda, and British Guiana, which at that time, the paper observed, would "all use low power to relay war news locally" (W.T. Arms, "Short-Wave News from Overseas", *New York Times*, 10 November 1940, 158). Berg (2013, 266) pointed out that American Armed Forces Services broadcasts and BBC programming had already become important in the Caribbean by 1944, in which year:

> ZQI, Kingston, Jamaica, 4700 kc., was on the air at 1730-1930, mainly in English.... AFRS [(US) Armed Forces Radio Services] transcriptions were broadcast regularly. From the same island, Cable & Wireless was sending out some BBC and American broadcast programming in the afternoons over VRR6, Stony Hill, on 15620 kc. And the Cable & Wireless station in Bridgetown, Barbados could be heard occasionally with broadcast material over VPO, 11475 kc.

The emerging popularity of radio in the Caribbean and the increasing importance of radio broadcasts to public audiences prompted major world powers to focus on broadcast operations in and to the region. These factors also resulted in international collaboration toward establishing and supporting Caribbean radio-broadcast presence. In 1946, the *Wall Street Journal* reported on an interactive broadcast prompted by global forces and featuring cooperation between two major allied governments:

> Uncle Sam and John Bull have been partners in a radio broadcasting business for three years. Every day since December 1, 1942, they've been on the air with a most unusual program – beamed toward neither U.S. nor U.K. listeners, but toward the myriad fold of the Caribbean. It was a war project at first, now there's a move to make it more permanent... The half-hour program, highly informal, is called "The West Indian Radio Newspaper." Its sponsor is the Anglo-American Caribbean Commission, an intergovernmental agency. From Washington the entertainment is piped by wire through New York to Boston where it is shot into the ether by two transmitters, WRUL and WRUW, whose direction antennae blanket the Caribbean. (A.K. Estill, "US–British Radio: Two Nations Broadcast Newspaper Every Day to Caribbean Neighbors", 10 January 1946, 1)

The audience for this programming included an estimated 18 million listeners, including some 5 million in US and British territories, including Trinidad and British Guiana. These figures assumed some level of communal listening over loudspeakers and other systems, and the interactive nature of the programming was evident:

> From the beginning the venture has been largely a good-will investment. It had some specific wartime objectives, such as telling West Indians how to eat when their fish supplies from Newfoundland and Canada were cut off – how to grow victory gardens, how to consume powdered eggs from the U.S., how to make native cassavas, eddoes and tannias into sprightlier new dishes . . . Listeners wrote from places like Lodge Village, British Guiana, asking what to do about sick rabbits and sheep. (Ibid.)

US Armed Forces radio broadcasts also added to the mix of global radio available in the region at the time. Holly Betaudier (2010), known primarily as a television personality in Trinidad and Tobago, wrote of his role in radio during World War II in Trinidad: "I was comfortable in 'radioland' with a programme, 'Holly's Happy Moments', that I started in 1946 on the US Armed Forces radio service network – WVDI in Fort Reid, Trinidad. At that time, WVDI mainly serviced the armed forces throughout the Caribbean."

These British territories, however, would have been primarily under the influence of the BBC. Specific focus on the islands started as early as 1939, with a programme entitled *Calling the West Indies*, which featured "West Indian troops on active service reading letters on air to their families back home in the Islands" (Newton 2008, 1). However, Sanders (1978) argued that it was not until after the end of World War II that the British authorities would realize the importance of broadcasting in the colonies to aid in reconstruction and to combat rumours and misinformation that were being spread in remote corners of the empire, and promoted the establishment of regular programmes of broadcasting. Sanders (ibid., 21) noted that in the case of Guyana, the British colonial government granted official permission for a private commercial radio station in January 1950, after the British Guiana United Broadcasting Company had become a subsidiary of the British Rediffusion Company, and on terms that were clearly in the interest of the British authorities: "The agreement was granted under the Post and Telegraph Ordinance and was tailored to the needs of the British Colonial Government. It included, for instance, a provision making it compulsory for the station to broadcast not less than twenty-one hours per week of the BBC's programmes. It also demanded ten and a half hours of air-time for the Government."

The initial reliance on British, American and even Canadian capital and expertise would be replaced by an emphasis on national control of mass-media resources in the period following the collapse of the West Indian Federation and the granting of independence to the major territories, when several regional governments "nationalized" the local electronic media. In Guyana, for example, the government acquired Radio Demerara from (the British) Broadcast Relay Services (Overseas) and combined it with the Guyana Broadcasting Service to create the Guyana Broadcasting Corporation (Dunn 2004a, 76). Similarly, an article in the (Canadian) *Ottawa Citizen* reported on a broader and clearly articulated policy of patrimonial media control in (then independent) Trinidad and Tobago:

> The Trinidad and Tobago government is taking steps to prevent foreign companies from controlling its newspapers, radio or television stations. In a proposed five-year development plan published Sunday, the government said it would acquire complete ownership of Radio Guardian as well as the controlling interests in the country's television service. Radio Guardian is one of Trinidad's two radio stations. It is owned by the Trinidad Publishing Co., which is controlled by Canadian-born newspaper magnate Lord Thomson. ("Trinidad Acts to Keep Control of Mass Media", 30 December 1968, 4)

An important part of this movement to nationalize media resources after independence was the notion (popular at the time especially among developing nations) that media represented a kind of natural resource that should best be used in the interest of national identity and social cohesion. In the cooperative socialist republic of Guyana, this relationship took on special importance amid concerns about racial tensions between the Afro- and Indo-Guyanese communities. Sanders (1978, 41–42) described how the allocation of particular slots of East Indian programming appeared to endanger national cohesion:

> There is one feature of the effect that commercial support for radio has on programming which is worthy of particular attention, given Guyana's history of racial tension. Both radio stations in Guyana have allotted specific broadcast times for "East Indian" programming. By this they mean that at specific times only music originating from India (usually connected with East Indian films) is played on the air. In Guyana, the programme-slots for East Indian music have always been regarded by the East Indians as "our" time and by the Africans, particularly, as "their" time.

Sanders (ibid.) reported over seventeen hours of Indian music and religious programmes per week across both Guyanese radio stations in 1977 and argued that the separation of Indian programming into discrete programming blocks was perceived as divisive. This was a problem that many observers thought could be simply solved by disbanding the Indian programmes and including the same content dispersed throughout the broader programming schedule. However, as Sanders (ibid., 43) further notes:

> GBS attempted, in 1976, to disband its East Indian programmes and to incorporate East Indian music within its broad format. The idea was welcomed by a majority of Guyanese, confirmed in a telephone survey conducted by the station. However, the mainly East Indian advertisers who paid for these programmes withdrew their advertising support from the station entirely and the station was thus forced to reinstitute the programmes as it could not find the revenue elsewhere.

Kamaluddin Mohammed and the Origins of Indian Content

While the present work focuses on Indian-format radio content in Trinidad and Tobago, it is important to note that such stations are a much later development in the history of media development in the region. Dedicated Indian-format radio stations did not emerge until the 1990s, while the history of Indian content on the radio in the region dates back more than fifty years prior. Thus it is necessary to pay some attention to the development of this prior form of Indian content before exploring its modern evolution.

For many, the history of Indian radio in Trinidad and Tobago begins with Kamaluddin Mohammed (Ghany 1996; R. Mohammed 2015; Parasram 2008). Kamaluddin, also known as "Chaarch" ("Uncle") or "Kamal Cha Cha" (Uncle Kamal) or just "Kamal", was born on 19 April 1927 to Fazal Mohammed and Khajiman Kartoum. He was one generation removed from indentureship, as his parents were both children of indentured labourers from India. His family was an influential one from an area known as El Socorro, along what is now called the East-West Corridor, but which at his birth would have been notable for its proximity to the railway and to rich agricultural plots. His early life included employment as a labourer and as a clerk some distance from home, in South Trinidad, after which he returned to the El Socorro/San Juan area ("Kamaluddin Mohammed Papers SC114", 27 April 2015).

Kamal would become a national leader later in life and eventually an international diplomat. He was a founding member of the PNM (notably, a predominantly Afro-Trinidadian political party throughout its history). For Kamal, his political and cultural agendas were closely aligned. In a 2011 speech Kamal said:

> My involvement in the formation of the People's National Movement in 1956 in Trinidad and Tobago with Dr Williams came at a time when I was simultaneously leading a struggle for the recognition of East Indian culture and music in this society. That struggle was one that tried to break the underdevelopment that was already a part of the colonial legacy of division between racial and ethnic groups in our society. (K. Mohammed 2011, 14)

Kamal reflected numerous global influences even at an early age, in part owing to the persistence of Indian and Islamic cultural influences from indentureship, but also to his own efforts. Parasram (2008) noted that Kamal was "versed in Islamic teachings and was fluent in Arabic, Hindi, Farsi and Urdu" and that "by 1947, at age 20, he became Imam at the Mosque at Queen Street, Port of Spain, after impressing skeptics with his brilliance as a theologian".

The fact that Kamal spoke multiple languages was notable because of his particular mastery of several linguistic forms, though it was also partly indicative of the evolution of the Indian presence in Trinidad. Even the use of the term "Hindi" in this context can be somewhat problematic, since it encompassed a continuum of linguistic forms ranging from standard Hindi to locally evolved and adapted linguistic systems. Mahabir (1999, 14) noted that the Indian languages spoken by the indentured immigrants evolved through the process of transplantation, social mixing and linguistic adaptation: "The majority of Indian immigrants were native speakers of various dialects of Indian Bhojpuri, but these were not particularly homogenous. In the situation a *koine* (lingua franca) variety of Bhojpuri developed in the sugar estates in Trinidad which was referred to as 'plantation Hindustani'."

As noted above, Mahabir (1999) also referred to the historical record indicating that the indentured labourers spoke several different languages, depending on their area of origin in India. These included Bengali and Tamil, which are so distinct that the new arrivals resorted to pointing to basic items such as rice and water and inquiring as to the words for these in their compatriots' languages. Added to this was the influence of English, French and Spanish that these labourers encountered in the Caribbean colony. However, as Mahabir (ibid.) also noted, the indentured labourers were also proficient in the reading and recitation of their religious scriptures in Hindi.

Given the several forms and different languages involved, there were clearly numerous global influences on Kamal and others in his community. These linguistic influences can be traced to the historical transplantation from India, which would account for Hindi and other Indian languages, and to the earlier influence of Islam in India, which brought Arabic and Farsi to the subcontinent and prompted the development of Urdu, which includes standard Hindi words and is written in an Arabic-based script. The particular expression of Islam in which Kamal was versed was also heavily influenced by an Indian experience, with Arabic being the formal language of scripture, but Urdu being the language of preference for religious literature and devotional songs (known locally as *Qaseedas*). Kassim (2002) noted that several schools had been set up in Trinidad during the 1920s and 1930s that taught Arabic, Urdu and Islamic religion.

Kamal's proficiency in translation among these languages and English, along with his religious knowledge, prompted his presence at the launch of the first broadcast radio station in Trinidad – Radio Trinidad – on 31 August 1947. The launch ceremony featured representatives from major religious groups in the island. Kamal was asked to arrange for blessings from the Hindu and Muslim faiths and it fell to him to translate the contribution of the Muslim representative (R. Mohammed 2015; Parasram 2008). Several sources suggest that his translation performance prompted the station to invite him to host an Indian music programme. Parasram (2008), for example, wrote that

> the Muslim representative at the blessing asked Kamal to translate the Arabic and Urdu blessings to English, a performance that so impressed the station's managers that they invited him to produce and present a show for the Indo-Trinidadian community. That was the birth of "Indian Talent on Parade", the radio show that was the first giant step in national recognition for Indian broadcasting in Trinidad and Tobago.

Diethrich (2004, 101) repeated this notion of station management's inviting Kamal to host an Indian music programme: "State-owned Radio Trinidad, the island's first radio station, was established on September 26, 1947. In its first week on the air, Kamal Mohammed – later to become one of Trinidad's greatest cultural impresarios as well a founding member of the PNM – was invited to host a program of Indian music."

However, it is important to note that this niche for Indian content was not necessarily the result of a sudden whim of station management based on Kamal's linguistic proficiency, or solely at the behest of the colonial owners and operators of the station. Outside of the accepted narrative of this sudden invitation to host an Indian music programme, there were, in fact, global forces at play in much more subtle and gradual ways. In particular, Kamal had been listening to a programme of Indian music on radio for at least several years from what was then British Guiana (today Guyana) and even struck up a correspondence with the host for some time before the launch of broadcast radio in Trinidad.

Among the global influences of the time in Trinidad it is also necessary to note the introduction of Indian films some twelve years prior to the opening of Radio Trinidad. The first of these films, *Bala Joban,* arrived in 1935, complete with songs and music that sparked a sense of global connection among the transplanted Indians (Niranjana 2006). Sharda Patasar (2014) noted that "while the silver screen was that lining of hope that allowed Indians in Trinidad during the early days an escape from pedestrian lives and hard physical work even while they saw familiar scenes and cultural similarities, the music of films lived on in the daily lives of the people". The lining of hope and the escapism of Hindi films for Indo-Trinidadian audiences notwithstanding, others have attributed a more substantial role to these films. Manuel (1997–98, 22), for example, suggested that these Hindi films did more than provide escapist reprieves for Caribbean audiences, serving also as one of the few connections to India: "The advent of commercial Hindi films to the Caribbean in the mid-1930s added a new dimension of Indian cultural presence in the diaspora. By the early 1940s Hindi cinema had become widely popular among Indo-Caribbeans, providing what many have perceived as a direct link to the cherished but otherwise remote homeland. Many Indo-Caribbeans value Hindi films as much for their Indianness as for their intrinsic features." Gooptar (2013b, 147) has argued that Indian films even influenced the choices of names for children of Indian descent in Trinidad, breaking with the traditional name choices and naming rituals based on astrology and astrological signs (*rashi*), writing that

after 1935 when Indian movies became popular in Trinidad and the Caribbean some parents began to move away from the time-honoured *patra* (horoscope reading) derived names for their children and began naming their children after their favourite male or female Indian film stars such as Raj Kapoor, Nargis, Premnath, Dev Anand and Hema Malini. These names were non-caste and popular with the diaspora. The practice continues today and there is a compromise by which many children are now given three names: the calling name, which may be referenced to one of the Indian stars or to some aspect of Bollywood in general, a *rashi* name and surname.

Yet, as powerful as the influence of films may have been, they did not always serve the cultural maintenance function that might be presumed. Jha (1973, 35–36), for example, placed the blame for the destruction of traditional Indian forms squarely on Indian films:

> Indrasabha, Raja Harishchandra, Gopichand and Sarwarneer were very popular dance dramas at one time among Indians of Trinidad. So were the Ramaleela, Krishnaleela and Rasmandal dance dramas. Now Indian films and their songs have almost killed these traditional shows. What one finds on the television, radio and social gatherings is an imitation of popular Hindi film singers' renderings. Now there are local Rafi, Lata, Mukesh and others – those whose voice resembles the artistes of those names from Bombay.

Niranjana (2006, 170) similarly argued that "once Hindi films and their music appeared in Trinidad, older folk music, classical music, religious music, and songs from the dance dramas either slowly disappeared or became imbued with the rhythms and instrumentations of the film songs", describing the process in the following terms: "During the 'cooking night' for Hindu weddings, participants enthusiastically sang Hindi film songs alongside the chutney lyrics common to such occasions. . . . The dance dramas popular until the 1950s were overtaken by the 'cabaret dance' of Hindi movie vamps and, more generally, 'film dances' performed to the film songs."

Mungal Patasar (1998, 70) similarly explained that the commercial Bollywood Indian music and dance forms served to displace well-established traditional forms, sometime through syncretic effects in which the more popular film forms came to be mixed with the fading traditional genres:

> During this period 1940 to 1960 [there] started a shift in the taste and aesthetics in Indian music. Form in the local classical singing remained. Singers like Sohan Gildharie, Raniff Mohammed and others maintained the form of the *thumarie* and the *ghazals*. But what started to happen was that the form of the film songs started to impinge on the *bhajans* and other religious songs. Today many of the film *bhajans* like "*banwarie re*" or "*jai, jai, jai tripararie*" have become part of the repertoire of

the temples in Trinidad. Hindus and Muslims use the melodies of the film songs to carry the message of their lyrics.

While Manuel (2000a) pointed to the later establishment of radio programmes focusing on so-called local or local classical music from Trinidad, this did not result in any meaningful restoration of these traditional forms (though they arguably contributed to the maintenance of these traditions and their eventual integration into later forms such as chutney). Broadly speaking, relatively modern Indian media eventually merged with (and often displaced) traditional cultural forms among the Trinidad Indians. The fact that these imports would then be held as elements of cultural identity points to both the global nature of cultural expression and the dynamic manner in which notions of culture may be constructed. As Sharda Patasar (2014) noted:

> Hindi film music was the epitome of "good Indian music" for that generation of Indians gradually becoming educated and city bred. There was a group of Indo-Trinidadians too who were exposed to popular American music of the 1930s and 40s, the music of calypso and Carnival which borrowed from Afro-American ragtime. The Bombay film music with its Indian-Western musical structure held similarities with sounds of the island while also new to the musical aesthetic of Trinidad.

Furthermore, the complexity of cultural morphology often results in transplanted cultural forms that differ greatly from their sources. This was evident in the fact that All-India Radio (to which Indians in Trinidad and Guiana were listening) refused to broadcast film songs after Indian independence, considering them to be insufficiently representative of Indian culture and values (Niranjana 2006).

Kamal Listens to BG Indian Radio

Manuel (2000a, 55) placed Indian content on radio in Guyana during the 1940s using both broadcast and the "Rediffusion" closed-circuit wired transmission system. However, Sidney's mention of Indian musical entertainment points to the fact that Guyanese efforts at Indian content on radio trace at least back to 1937 with a programme called *The Indian Hour of National Music*, described as "a commercial programme broadcasting Indian music and songs which, by 1946 was run every Thursday 8.00 pm local time" (Akbar 1946) on station ZFY. The host of this programme, Mohammad Akbar, wrote to Kamal on 11

January 1946 to acknowledge Kamal's letter (written in Urdu) and song request. Akbar's response was short but cordial and indicated that the cost for a song request was $1.50.

However, this was not to be the end of their exchanges. Kamal and Akbar continued to correspond for some time and, beyond adding a regional dimension to Indian music radio programming, their letters reveal that the start of Indian radio programming in Trinidad most likely did not originate solely with an invitation from the radio-station management and their being impressed with Kamal's translation skills. The evidence suggests that Kamal had discussions with Akbar in the weeks preceding the launch ceremony about hosting such a programme on Radio Trinidad. A letter from Akbar dated 30 August 1947, for example, read in part: "I am in receipt of your letter dated 21st Aug. and wish to congratulate you on your venture to run an 'Indian Programme'. And trust that you will be very successful in your undertaking" (Akbar 1947). Akbar commented on Kamal's proposed rates for advertising and requests being very high, gave him advice on presenting and managing the programme and promised: "I shall give you a spot on my programme on Thursday, 4th Sept, intimating listeners to tune in to your programme on Sunday, 7th Sept." Though subsequent letters indicate that Akbar missed Kamal's first broadcast, he did eventually listen to the programme from Trinidad and, having listened, gave Kamal feedback on the quality of the signal and the content of the broadcast. He suggested, for example, that the presence of famous dancer Champa Devi was a welcome addition to the programme and that she should be featured more prominently.

While the name Champa Devi was no longer necessarily associated primarily with radio among participants in the present fieldwork, her influence on the broader context of musical expression in the Indo-Caribbean context bears some mention. Manuel (2000a, 172) briefly mentioned her as a "flamboyant Trinidadian dancer of the mid-century decades" and notes that she was among those who challenged the performance roles for women in song and dance among Indo-Caribbean communities. Bergman (2008) has, in part, chronicled Devi's rise to popularity and her influence on both the Indo-Trinidadian dance and music of the time, starting about a decade after the introduction of Indian films. Additionally, the fact that this (notably) female star was an important element of the early Indian programming on Trinidadian radio and Akbar's advice to feature her more prominently cannot be easily separated from her regional appeal. Bergman (ibid., 9) wrote of Champa Devi: "Beginning dance at age 15, she first became famous in Suriname and Guyana before returning to Trinidad to perform as the star of a stage production called Gulshan Bahar in 1943. . . . Champa Devi was largely self-taught, her dances inspired by Indian films."

While Akbar, with his advice on presenting and on content, acted as a distant mentor, another listener from Guyana would become important to Kamal, in the person of Azeem Khan, who first wrote to him on 9 September 1947 indicating that he listened to *"Indian Talent on Parade"* and expressed his "desire to report that reception was excellent" (Azeem Khan 1947). Khan, who also became a broadcaster, offered to help with the efforts of broadcasting Indian programmes. Kamal would remember these men and the contribution of Guyanese Indian radio broadcasts many decades later, when he recalled the circumstances in which Trinidad audiences would depend on shortwave broadcasts from Guyana:

> In those days, before Radio Trinidad launched my programme in 1947, those who were fortunate to have radios and who were inclined to listen to East Indian music were only able to listen to it on Radio ZFY from British Guiana. The reception was not always clear and people in Trinidad and Tobago came to know the names of those British Guianese radio announcers such as Azeem Khan, Mohammed Ackbar [*sic*] and Dindial Singh. (K. Mohammed 2011, 14)

While radio was already an established communications medium during the 1940s, it did require investments in technology (particularly prior to the introduction of transistor radios) that were substantial for many in the Indo-Trinidadian community. Not all families could afford a radio at the time when Kamal started broadcasting Indian content, a problem solved in part by communal listening. Mungal Patasar (1998) described Kamal's pioneering radio efforts and the often-communal enthusiasm for this content in the context of the emerging popularity of Hindi films and modern Indian film music:

> From 1940 things started to accelerate. Indian films were introduced to Trinidad. The film songs had a great impact on the music of Trinidad. Kamalaudin [*sic*] Mohammed also got his first radio programme during this period, and every Friday evening at 4.30 p.m., all villagers would put on their radios; those who could not afford a radio went to a friend, and even if they had to stand outside, they listened to the radio programme.
>
> Kamal's voice was like music to the ears of every Indian. There was a revolution in music listening (70).

For Parasram (2008), the introduction of Indian radio programming in Trinidad was meaningful for several reasons, including the assertion of identity among the transplanted population, which by this time included many members who were born in the new homeland:

> In 1947, Indians were still a minority in colonial Trinidad and Tobago, with a culture that was generally unknown and misunderstood. Indians were often the objects of derision because of their language, religion and culture. "Indian Talent on Parade" became the nation's first mass media vehicle for the Indians. And Kamal used it to begin creating an understanding and appreciation of the Indian community, its art and culture, as well as its religions. . . . For the Indians, it was a major step forward, giving them self-confidence and the respectability they deserved. . . . The impact in 1947 was astounding as he began what was in effect an experiment in ethnic broadcasting.

This experiment in ethnic broadcasting, spearheaded by a figure who, as a Muslim, was himself a minority figure within the Indo-Trinidadian ethnic minority, allowed for the creation (or imagination) of conceptual linkages between the diaspora members and their homeland fitting with Ramnarine's (2011) circulation of creative practices. Sharda Patasar (2014) emphasized the linking function of the advent of Indian radio content and suggests a global dimension, writing that "images of Indians sitting around the radio in the village shop to listen to Kamaluddin Mohammed's half hour programme, 'Indian Talent on Parade' in the late 1940's speak of the importance of music as a link to a culture from which Indians had been geographically separated".

More than six decades later, Kamaluddin Mohammed (2011, 14) would reflect on the launch of Indian radio content in similar terms:

> My involvement with the launch of the programme "Indian Talent on Parade" in 1947 on Radio Trinidad was not part of any desire to create separation in this society. What we were dealing with was the under-representation of the cultural identity of a very large sector of the colonial society in Trinidad and Tobago. There were those who felt that it was an attempt to promote separation, while others recognized that it was all about validation of cultural identity.

Regional forces and cultural exchanges were also evident in the evolution of these local media forms. One of the first performers on Kamal's *Indian Talent on Parade* programme was a singer named Ganga Persad. In a tribute to Persad on his death in 2010, Seeta Persad in *Newsday* ("Singing Legend Ganga

Persad Dies", 11 October 2010) noted that the programme propelled the singer to stardom not just in Trinidad but also regionally: "He was requested to sing at many religious-based celebration as well as weddings and fetes. He was invited to perform in Guyana and Suriname, where those who heard him remarked they felt they were in India. In Trinidad, the enthusiasm and support for Persad was no different."

Conclusion

The emergence of radio in the region provided a new window on the world. Set up primarily as a means of securing imperial influence through direct communication, this medium quickly caught the attention of the creole populations, including the indentured Indians and their descendants. Earlier efforts at programming aimed at the Indian indentured labourers and their children in Guyana served as a starting point and a model for Kamaluddin Mohammed, who distinguished himself by creating such content on the radio in Trinidad, using a combination of Indian and Trinidadian material. This small start would furnish the basis upon which future developments would be built – including such gains as regular slots in mainstream stations and, eventually, Indian-format stations.

CHAPTER 4

Media and Indian Content in Trinidad and Tobago

As mass media evolved in the region and in Trinidad and Tobago, the so-called creole mainstream folklore, along with colonial norms, tended to dominate content. Content aimed at the Indo-Trinidadian community or content that addressed its concerns was rare in such a media environment, prompting the establishment of an ethnic press. Later, when electronic media became available in the form of radio, limited slots of time were made available for content aimed at that community. Over many years, the local industry resisted the notion that the market could support anything more than these dedicated time slots and they became established as the media voices of the Indo-Trinidadian community alongside alternative media forms such as local megaphone players. This evolution of local media and the role of Indian content prior to the establishment of the first Indo-Trinidadian station are examined in this chapter.

The Colonial Era

Mass media in Trinidad and Tobago date back to the 1820s, when the *Port-of-Spain Gazette* newspaper reported on events in the colony, in the United Kingdom, her other colonies and further afield. That particular paper came to an end in 1956, but several other major dailies evolved and flourished in the early colonial period. Another newspaper, the *Trinidad Chronicle*, started in the colonial period (circa 1864) but went out of circulation in 1959 prior to independence. Other major titles, some changing their mastheads and management over time, included the *Daily Bulletin* (starting as early as the 1880s and running into the early 1900s) and the *Mirror* (from the late 1890s up to 1932), which then circulated as the *Daily Mirror* up to 1967. In colonial Trinidad, several parochial/ethnic publications were also popular, including titles such as the *Catholic News,* dating from 1892, and the *Indian Weekly,* which was produced in the 1920s and 1930s.

In *Finding a Place: IndoTrinidadian Literature,* Kris Rampersad (2002) has argued that the very exclusion of the Indo-Trinidadian voice from the mainstream colonial press was the driving force behind the emergence of several Indo-Trinidadian print publications in the colonial era, arguing (ibid., 7) that "as early as the 1870s IndoTrinidadians were very much involved in attempts to articulate a national ideology, notions of national identity and the place of Indians in Trinidad society". Rampersad thus placed the origins of Indo-Trinidadian media with nineteenth-century Indo-Trinidadian letter-writers to the *San Fernando* and *Port-of-Spain Gazette* newspapers who faced resistance to having their ideas published, leading to the establishment of special-interest ethnic newspapers such as the *Koh-i-noor* (*Indian Koh-i-noor Gazette*) on 4 October 1898 (ibid., 82). Later, the community would also produce other publications, including titles such as the *East Indian Herald* and the *East Indian Patriot.*

As we have seen above, both wired and broadcast radio started in the colonial period, with emphases on the needs of the colonial representatives being their primary focus and content for the locals being of secondary importance. Potter (2008) and Wilson-Heath (1986) are among those who clearly point to the idea that British broadcasting efforts, whether to the colonies from Britain through the "Empire Broadcasting" service or in the colonies themselves as stations, were established with the stated objective of serving "the white population under the British flag" (Potter 2008, 475). Yet locals and non-white presenters were allowed time on the air, and programmes that interested local populations were broadcast in limited amounts on stations established in various colonies. This was the case in both Guyana and in Trinidad and Tobago with reference to Indian content catering to the ancestral Indo-Guyanese and Indo-Trinidadian populations.

Independence, Postcolonial Media and the Nationalist Turn

The (some would say pragmatic) divestment of British colonies following World War II and similar moves by other European colonial powers coincided with an increase in international diffusion of the technologies of mass media. Many of these ex-colonies would see the introduction or acceleration of various relatively new technologies along with their new independence. In Trinidad and Tobago, television was introduced around the time the country gained independence from the United Kingdom, though radio had been popular for many years before, and several newspapers already existed.

A relatively common theme among newly emerging nations of the time, and one discussed at international venues such as the United Nations, was the question of how mass media, perceived as powerful tools of influence, could be used in the mission of building new nations through informing their populations and promoting national aspirations and identity. Wilbur Schramm (1964, 114) wrote that "mass media are essentially agents of social change . . . such changes in behaviour must necessarily present substantial alterations in attitudes, beliefs, skills and social norms".

Cuthbert (1977, 90) noted that mass media in developing countries play a different role than in the developed world, such that the rapid rate of change in developing countries made it "extremely important that the means of communication be purposefully used in the interests of the nation". This notion of "purposeful use" found favour with many leaders and policymakers in developing and newly independent nations. Media were run with a variety of state influences, from outright ownership and control to licensing and monitoring. Most television was at that time generally too expensive to be set up with private resources in developing nations, and stations existed only as state enterprises and often as tools of government information. At the same time, these emerging societies began to take notice of the influence of foreign content on their nations. With the departure of formal colonial authorities, questions were being raised about the influence of foreign media as a kind of electronic colonialism in which foreign values and ideas were quietly influencing their national culture. Scholars such as Schiller (1976) and McPhail (1981) explicated these perspectives under the banner of cultural domination and electronic colonialism.

To be clear, this debate was not solely an academic one, nor was it limited to mass media scholars. President of the Caribbean Development Bank William Demas (1975, 4) complained that what he called "cultural penetration" (a term widely used later by others) by foreign mass media was destructive of not only national cultural identity but also of "autonomous and independent economic and social development" arising "not only from the advertising of imported goods but also from the actual content of the programmes which brainwash the population into accepting and wanting the way of life of the affluent societies".

As Merrill (1971) wrote, a wide range of communications scholars had, by the 1970s, established ample data from a multitude of studies that showed correlations between the use of mass media and the achievement of various development goals, including economic growth. At international meetings such as those of the United Nations, the newly independent former colonies and other developing nations took to advocating what would eventually be known as the New World Information and Communication Order.

As with many of its newly independent counterparts around the world, Trinidad and Tobago wrestled with the roles and functions of mass media in its emerging post-colonial society. There were some early efforts towards nationalization of media resources in the post-independence era, partly due to nationalistic tendencies but also due to the prohibitive costs of new technologies, particularly television. There was, however, never a full widespread nationalization of media resources or official control of media such as was attempted in Guyana to the south. After independence, the mass media in Trinidad and Tobago consisted of a few distinct operations, including both public and private media.

The government-owned and controlled television station known as Trinidad and Tobago Television (TTT) launched on 31 August 1962. Despite government ownership and repeated accusations of governmental interference, TTT continued to claim independence from politics. TTT monopolized television broadcasting in Trinidad and Tobago from independence until the introduction of competition in the years following the attempted coup of 1990. This competition came in the form of two rival stations under private ownership, the small operation known as AVM (Audio Visual Media) TV4 television and a much larger and successful station known as CCN (Caribbean Communications Network), TV6, both launched in 1991. TTT went off the air in early 2005, when it was replaced by a new government-owned entity. By that time, however, the station had undergone several management changes, rebranding, and had been merged with government-owned radio services over more than a decade prior.

In later years, the introduction of cable television would drastically change the availability of television viewing options. Several independent operators began to appear throughout the island in the late 1990s and early 2000s, with a merger creating the Cable Company of Trinidad and Tobago, which was eventually acquired by a multinational operation named Columbus Communications in 2006. Several other providers have since emerged, delivering cable television, phone and Internet services. The spread of cable gave rise to the emergence of several local cable-only television special-interest broadcasters, including Indian content stations as well as religious stations. However, the majority of cable content (as well as the content of free-to-air television) has been dominated by US television programming, which accounts for as much as 75 per cent of content by some estimates (McFarlane-Alvarez 2007). This level of content from the United States has been a matter of some concern for several decades among Caribbean scholars and media practitioners (Brown and Sanatan 1987).

In the post-independence period, the two radio stations in existence were the privately owned Radio Trinidad (730 AM), which had its start in 1947 under colonial rule, and the government-acquired National Broadcasting Service (NBS) Radio 610 AM. These stations eventually diversified into FM broadcasting and launched music-broadcasting services on the FM band during the 1970s. Liberalization of the radio market in the early 1990s saw a flood of new privately owned stations that covered a wide spectrum of music and entertainment formats, including adult contemporary, local soca and regional reggae, sports, talk and Indian music. The Telecommunications Authority of Trinidad and Tobago listed a total of thirty-eight broadcast radio stations in Trinidad and Tobago in 2015 with combined gross revenues of approximately TT$200 million (about US$29.5 million), roughly 10 per cent of the total broadcast market revenue of TT$1.08 billion (about US$160 million), which also includes broadcast and cable television (TATT 2016).

Three major daily newspapers dominate the modern media market, namely the *Trinidad Express,* which started in 1967, the *Trinidad Guardian,* dating back to colonial times in 1917, and the more recent addition, the *T&T Newsday,* which started in 1993. At independence, the *Trinidad Guardian* and the *Daily*

Mirror were both in operation as British-owned enterprises and the *Trinidad Express* emerged as an indigenous competitor. Later, the *Daily Mirror* would cease to operate, and local business interests acquired the *Trinidad Guardian*. At present all three of the local major dailies are locally owned and operated. In addition to the major dailies, several weekly publications remain popular. These include titles such as the *Bomb* (which started in the early 1970s), the *TnT Mirror* (from the early 1980s), and the *Sunday Punch*.

Though the present work focuses on Indian-format radio in Trinidad and Tobago, it is necessary to point out that this particular media sector was not the only avenue for Indo-Trinidadians' involvement in local electronic media. As electronic media evolved in post-independence Trinidad and Tobago, Indo-Trinidadians became part of the industry mainstream. Television personalities such as Yusuff Ali, Jai Parasram and Salisha Ali were part of the news-anchor lineup, along with others in current-affairs programming such as Dale Kolasingh and Tony Deyal. Even in entertainment programming, figures such as artist Ian Ali were evident. As we shall examine later, Hans Hanoomansingh was part of mainstream radio, as were Sam Ghany and several others.

By 1995, political changes began to focus attention on the role of Indo-Trinidadians in the media. As veteran journalist Afzal Khan had earlier come to head the news department of the sole television station, and other Indo-Trinidadians such as Gideon Hanoomansingh (later to become a politician) were increasingly visible as news and current-affairs personalities, questions were raised on the one hand about whether the ethnic group was adequately represented in the media, and on the other, whether their increasing media roles were reflective of political interference. Evidence of this debate can be found not only in press reports and letters to newspaper editors, but also in the publication of a 1995 research report from the University of the West Indies Centre for Ethnic Studies called "Ethnicity and the Media in Trinidad and Tobago" (sometimes also called the Ryan Report).

Indian Music on the Radio

For several decades following the launch of Radio Trinidad and the early contributions of Kamaluddin Mohammed, the presence of Indian music on radio (and, later, television) was limited to particular programming slots within the broader mainstream programming. Interviewee Kenny (2016), from the central town of Couva, who described himself as being "nearing sixty", recounted, for example, that

> when, like, Indian music came out like on Sunday mornings, from 6:00 to 6:30 there was only one show . . . and everybody listened to that. So that was like a favourite, then, on a Sunday morning, because English music was [played] right through all the time . . . So on Sunday morning you would get the 6:00 to 6:30 half an hour; only on 730 AM radio station that you get that half-an-hour programme.

The audience for Indian music of both local and foreign types thus found itself in a situation where the availability of such music on the radio was often far less than it would have liked. Several interviewees have told the story of people driving through certain villages and hearing the Indian programming emanating from the homes. One informant indicated that during the Sunday morning broadcasts, enthusiasts would amplify the radio programmes on loudspeaker systems to crowds shopping at outdoor markets such as the Tunapuna market.

Indeed, according to some of the older interviewees in the present research, in the decades following the popularization of Hindi film songs, Indian music (whether secular or religious in nature) was considered acceptable in many households, whereas English or Western music was considered to be morally unacceptable. This particular moral framework was enforced despite the fact that most of the younger listeners at that time had already lost the Hindi language and could only understand the lyrics of English songs.

Under this general understanding of the hierarchy of acceptability, the most sexually charged Hindi film songs, couched, as they often were, in oblique poetic and euphemistic portrayals of romantic love, were preferable to Western musical selections with direct attestations of emotion and references to physical acts such as kissing. Within this unspoken hierarchy as well, the emerging local creole calypso and soca forms were considered outside the pale and not allowed in many Indo-Trinidadian households, whether played on the radio or brought in as records. This broader notion of Indian content being somehow morally superior tends to persist today among older members of that community (and even in at least one interview with a staff member at an Indian radio station).

Paradoxically, when pressed on the question of their music-listening patterns in the past, some of these very same older collaborators expressed fond memories of the Beatles and Paul Anka from their younger days and even referred to their collections of Jim Reeves, Glen Campbell, Skeeter Davis and Charley Pride vinyl records. Both in formal interviews and in informal conversation, Hans Hanoomansingh also made the point that musical nostalgia, even among Indo-Trinidadians, necessarily includes popular American music that dominated airplay in years gone by. In some cases, as well, there was reference to some people having the occasional calypso or soca record in the home, starting in the 1970s, even if only as a concession to the immense airplay popularity and availability of such records. Among these, there was also some distinction made among acceptable calypso/soca performers and those known to use vulgar lyrics or who were commonly felt to denigrate the Indian population.

Under this hierarchy of acceptability and within the complex mix of musical forms in Trinidad, there existed a continuing gap for Indo-Trinidadian music fans in which the audience for Indian music, driven by its popularity at local concerts and Indian movie showings, only encountered a limited availability on the radio. Kenny (interview by author, 12 April 2016) recounted that during the 1960s and 1970s, when he was a young man, it was necessary to buy records to fill the gap in the availability of Indian music. While they were not expensive, the 78 rpm records, followed by singles and twelve-inch albums, represented a substantial investment in many cases and required careful storage and handling in an environment where direct sunlight was a constant threat. Kenny (ibid.) said: "You know, money had value. So, like, when you buy a record for, like, $2, $1.50, $2, $3, you pay for a record then. You know – like if you spend a $5 it was like $200 now."

For music enthusiasts like Kenny, growing up in the 1970s, the records came at some cost and often with limited availability (new releases might be sold out and older songs might not be stocked in the record stores). Other options for hearing Indian music included commercial concerts, where, for about five dollars a person, one could attend locations such as the Himalaya Club in Barataria and the Rienzi Complex in Couva to hear local classical artistes such as Isaac Yankaran and Sohan Gildhari. These concerts would, according to Kenny, run from about 5:00 or 6:00 p.m. and continue to 11:00 or midnight. On a more informal basis, the opportunity to hear Indian music also arose at the celebrations surrounding the Indian (mainly Hindu) weddings in Trinidad where one could hear both the local classical and Hindi film songs played. A key factor in the accessibility of this music for many decades has been a particular combination of technologies commonly known as the "mic" or "mike" system.

The Mic Men

A frequently overlooked communication technology and social phenomenon sometimes filled the gap between the desire for Indian music and the amount of that music that was available to listeners. The mic man (or mike man) was a standard feature in Indian communities in Trinidad. The mic man (and they have traditionally been only men – though this may be changing) was an individual who served as a kind of public communicator or modern town crier in most villages. Popular in rural areas, these mic men were also known to operate in suburban areas around the capital and other cities and continue to do so today.

To play his role of communicator, the mic man attaches one or more large (approximately twenty-two to twenty-four inches in diameter) public-address horn speakers to the top of a car. These speakers are connected to an amplifier inside the car. In the simplest systems the driver (or another person in the car) uses a microphone plugged into the amplifier to read messages to the community while driving through the area. These messages are usually paid announcements and, quite often, the loud tinny sound of the horns would actually announce the death of a person in the village. For this reason, and because the announcements would be made without precursor, the "mic" would draw considerable attention when heard and thus has been used for many other purposes, such as community announcements and in political campaigns (Robert Ramroop, interview by author, 29 May 2016).

However, in addition to its death-announcement and town-crier functions, the mic system also evolved the ability, with the addition of record players in the vehicle, to play music for audiences. Gooptar (2013a) traced the evolution of the mic systems back to the 1940s, and mic man Robert Ramroop (interview, 29 May 2016) confirmed this general time frame, though no one has suggested a more definitive date for its origins.

Gooptar (2013a) also posited another important role for the mic men in the development of Indian music and audiences in Trinidad, suggesting that these "mic" systems were initially employed as amplification systems for classical singing events, including "cooking-night" festivities, to enable audiences to hear the live performances more clearly. Ramroop (interview, 29 May 2016) also described this live-performance amplification as important in a variety of events (even today) and noted that in some early film showings, the mic served (for a brief period) as an amplification system so that the audience could hear the soundtrack. From these beginnings the mic became a mainstay of entertainment, including establishing its presence as a source of amplified recorded Indian music at various festivities.

Ramroop (ibid.), when asked to describe the origins and first uses of the mic system, said that he remembers it primarily used "to carry weddings", which he described as participation in the ritual processions to and from the homes of the bride and groom: "In Hindu weddings when you have to go down by the girl house and you need some music, you put two funnels on the car, you have your amp in the car and you going down with the wedding." From these uses, the "mic" also evolved a presence as a link in the chain of mass media marketing, becoming a medium for promoting new Indian film songs, sometimes even before the films were released. In this way, the mic men garnered attention and business for themselves while performing important marketing for the music and film distributors. According to Ramroop (ibid.):

> Mr Balroop . . . and a fella they called Razack from San Juan. They used to give you sample records and tell you to go out and play the songs. They used to give us, but it was a business for them. That was the tar records. They used to bring the first of the records. As tunes come out they used to have it . . . to advertise the films. . . The film opening next week, they want you to play the songs so people could hear what is in the film.

Ramroop (ibid.) recalled the records as coming from EMI and Gramophone in India and credits the mic men with providing music to their audiences in Trinidad, noting that the traditions of public announcements and wedding music continue with the mic men today, providing a source of hobby activity and income for at least 135 members in their association.

Radio Guardian

Radio Trinidad was the only broadcast station in Trinidad and Tobago for several years and for that time held a monopoly on the ability to broadcast radio and (however limited) Indian content. The opportunities to feature Indian music content would increase with the launch of the station known as Radio Guardian, owned by the British Thompson Group, in 1957, on the AM band at 610 kHz, or its acquisition by the Trinidad and Tobago government in 1968.

The introduction and development of Indian content on Radio Guardian was by no means automatic. The specialized programming included Sunday-morning programmes that would occupy full programme lengths of an hour or more and shorter slots, usually about fifteen minutes' duration, during early-morning and late-evening schedules during the week. Among the contributors to this programming and its growth was a young broadcaster by the name of Hans Hanoomansingh, who joined Radio Guardian (later called 610 Action Radio after it became part of the National Broadcasting Service) as a broad-caster in 1961. Hanoomansingh (interview by author, 15 July 2015), who now holds an honorary doctorate from the University of the West Indies for his work in culture and media, recalled his early involvement with Radio Guardian:

> What happened, when I became a broadcaster in 1961 – Nazim Muradali and I joined – it was called Radio Guardian. And Radio Guardian was part of the Trinidad Publishing Company – that included Trinidad Guardian and TTT. So, when we were receiving orientation to the new place on the first day that we joined, Nazim asked a question of the programme director: "Do you play Indian music?" and the answer was, "No, we do not play Indian music and we do not do sports coverage – live sports coverage."

Hanoomansingh, who has also served as an elected member of the Par-liament of Trinidad and Tobago, also expressed something of the nationalist sentiment that would have been consistent with broadcasting in the 1960s and concerns about foreign content (he uses the term "English music", most likely referring to American music, which was already dominant) and local relevance. However, Hanoomansingh included in that nationalism an implicit argument for Indian music as being relevant to Trinidad and Tobago, since it addressed the needs of a large segment of the local population. Hanoomansingh (ibid.) recounted his opportunity to influence the content of Radio Guardian to include Indian music as part of his service to Trinidad and Tobago within a few years of his joining the staff in 1961:

> The opportunity to change that and to make, as a consequence, the programming of Radio Guardian relevant to the reality of Trinidad and Tobago, presented itself to me. I wasn't an executive, I wasn't a programme director, I had no hand in programming. But I was able to influence the introduction of Indian music to Radio Guardian. . . . So I was able to make Radio Guardian relevant to Trinidad and Tobago. At that time, Radio Guardian could have been located in Los Angeles or Toronto or London,

because it was dominated by English music, and canned English music, so there was no relevance to what was happening locally.

Hanoomansingh's first programme of Indian music at Radio Guardian was a fifteen-minute slot that he called *Melodies of India*. The commercial success of this programme prompted station management (which Hanoomansingh notes was not Indo-Trinidadian) to extend the show to a half-hour. According to Hanoomansingh (interview, 15 July 2015), commercial success was a firm basis on which to ground additional programming such as his *From the Silver Screen* programme of Hindi film music, which started in 1964:

> And when we started that programme, it was film music from the "silver screen". There was a distributor of Indian films here, India Overseas International, and they supplied the music. They were the sponsors, together with the Kirpalani Group of Companies. The directors of India Overseas lived in the Woodbrook area, and on a Sunday afternoon, would be travelling to San Fernando, where they had two cinemas: Metro and Palace. And they were happy to tell me that when they entered Caroni – no highway at the time – the Old Southern Main Road – that they didn't need to have their car radio on, because every radio was playing *From the Silver Screen*.
>
> The first ever survey of radio listenership was held somewhere around 1967, and of all programmes on radio, of all programmes – calypso, rhythm and blues, whatever – *From the Silver Screen* commanded the largest audience – significantly more than any other musical programme on the radio.

Hanoomansingh (ibid.) pointed out that the second most popular programme was another of his Indian music offerings, called *Gems of India*, and that the commercial and popular success of these offerings were an indication of "the extent of which Indian music, on Radio Guardian, had fulfilled a need that was denied before *Melodies of India* came on the air". Besides *Melodies of India*, *From the Silver Screen* and *Gems of India*, Hanoomansingh also pioneered several other short programmes, including one titled *Cultural Traditions* and another called *Jhai Bharati* (*Indian Echoes*).

Despite the apparent profusion of these programmes, however, they represented only minor parts of the programming day, and were often scheduled outside the prime programming time slots. The reaction of audience members, however, was quite enthusiastic, according to Hanoomansingh (ibid.), who recalled their accounts even many years later about how important these programmes were perceived to be in their lives:

Many years later, I went to a temple in Debe and there was a function where I was asked to speak. And in introducing me, the chairman, Dr Nissir, told a story. He said he grew up in a place called Lengua. He attended Naparima College. His responsibility in the morning was to take the family's box cart and fetch water in the standpipe. And he would rush to do that to come back home to listen to a programme I did, a programme of Hindu music fifteen minutes long – I could only play three *bhajans*. But he said that was enough to energize him for the rest of the day. And that is how the country, the Hindu society, also responded. Fifteen minutes of music, at 5:45 in the morning.

Hanoomansingh, while celebrating the opportunities Radio Guardian provided for broadcasting Indian music and cultural content, did note that the music and culture were not always (and may still not be) considered part of the national cultural mainstream. Indian music and Indo-Trinidadian content were not only marginalized from the cultural mainstream, but also treated with some hostility. While specifically downplaying the incident, Hanoomansingh recounted an event in which another broadcaster at Radio Guardian deliberately scratched the Indian records so that they could not be played on air. He pointed out that this was the action of a single individual in the context of a station that was willing to allow Indian music on its schedule. However, many years later, the othering of Indian music and its exclusion from the cultural mainstream continue to be problematic in the broader social context as well. As Hanoomansingh (interview, 15 July 2015) noted: "If you travel by maxi taxi, or taxi, and even if the driver is Indian, you're not going to hear Indian music. . . . It's not likely to happen. It's not likely. Public transport will not play a station that provides exclusive Indian music or celebrates festivals of the Indian community." Pundit Jeewan Maharaj (interview by author, 15 June 2015) noted that despite the paucity of Indian music featured on media in the days before Indian radio, the little music that was played proved to be a source of contention:

In the old days you had to get up really early in the morning to catch an Indian programme. If you sleep late you'll miss it . . . and certain times in the evening time. And even then calypsoes were written complaining about how much Indian music there was. So you had a fella like (Lord) Shorty singing: "When you listen on the radio it's only Indian tunes and calypso." . . . There he was singing about how much Indian programmes they had. So some of us were saying we had a little bit, while others of us were of the view that "Anytime you listen on the radio, is only Indian tune you hearing."

Indian Music on Television

While Indian music and cultural content continued in small niche segments on radio in Trinidad and Tobago from the late 1940s onward until the launch of 103 FM in 1993, new developments in media brought new programming opportunities. These developments, however, were neither inevitable nor simply achieved. In August 1961, for example, *Variety* ("TV for Trinidad Gets a Setback", 9 August 1961, 20) reported that hopes for television in Trinidad had dimmed with the refusal of the US authorities in Washington of a proposal to establish a station at the US naval base at Chaguaramas (a radio station with call letters WVDI was being operated there). This proposal had formed part of negotiations among officials of the United States, the United Kingdom and local authorities, based on the revision of lease agreements for such bases, and had been greeted with initial support. The proposal envisioned that a small station would be established using US expertise and would be used to train locals to run the operation one day. However, US authorities eventually determined that the size of the base was too small to justify the expenditure that would be necessary.

Eventually, commercial arrangements worked where military and diplomatic ones would not. TTT started broadcasting in 1962 (Dunn 2004b; J. Rampersad 2012) on VHF channels 2 and 13. At the behest of local authorities, a foreign crew and management staff associated in part with US television giant Columbia Broadcast System initiated TTT's operations at what was known for many years as Television House at 11A Maraval Road in Port of Spain, while hiring and training local staff to take over the operations. In the month preceding establishment of TTT's regular broadcasts, *Stage and Television Today* ("Trinidad on the Air Next Month – 30,000 Homes by 1964", 18 October 1962, 10) reported its launch as follows:

A full television service will be operating in Trinidad by November 1st. . . . The shareholders in the company are Rediffusion (West Indies) Ltd, Scottish Television Ltd, The Columbia Broadcasting System Inc, and the Trinidad and Tobago Government. The potential audience in Trinidad is 100,000 homes, this being the number of houses which have an electricity supply. . . . The main transmitter operating on Channel 2 is located at the Montserrat Hills in the approximate centre of the island. The site is 1,000 feet above sea level and the transmission provides coverage over 95 per cent of the island. The studios and offices are in Television House in Port of Spain, the capital.

The challenges of furnishing programming for a full-time television station proved substantial, and the station operated for many years on a limited time basis, with a four-hour schedule starting at 6:00 p.m. The expenses associated with foreign programming and the emerging nationalist sentiment arising out of independence fuelled some attention to local production, including the flagship news programme *Panorama* and field-recorded magazine programmes such as Horace James's *Hibiscus Club*, as well as live productions such as *Community Dateline* and *Mainly for Women*. Another major thrust included talent-competition programming that attracted community involvement, sponsorship and viewership. Holly Betaudier, who left hosting at WDVI on the US base to try his hand at television, was both an on-air personality and a media salesperson (J. Rampersad 2012). Betaudier drew community participants, sponsors and viewers with a local talent competition known as *Scouting for Talent*. Other talent programmes included a long-running juvenile competition known as *Twelve and Under*. The talent-show model was also the basis for one of the few elements of Indian programming on TTT, a programme known as *Mastana Bahar* (loosely translated as something akin to "Joyful Season" but probably closer to "distant indulgence" or "intoxication from afar") that started in the early 1970s.

This programme was the mainstay of Indo-Trinidadian content for several decades in Trinidad and Tobago and the source of much contention as a visible marker of Indian cultural identity. One of the constant criticisms of the show, which were fuelled at least in part by Betaudier and others, was that sponsors unfairly dedicated money and resources to the Indian programme while neglecting other shows such as *Scouting for Talent*. *Mastana Bahar* did in fact receive sponsorship from major businesses associated with the Indian community – though critics never clearly established a basis for how this could be considered improper or unfair.

Manuel (2000b, 335–36) described the *Mastana Bahar* phenomenon as "an Indo-Trinidadian amateur song and dance competition network", noting that the show was "founded in 1971 by Sham and Moean Mohammad [*sic*], two energetic entrepreneurs also prominent as radio deejays, record producers, and concert impresarios" and "has evolved into an institution in Trinidadian culture". Manuel (ibid., 335) has, however, been less than complimentary of the actual content of the performances on the programme, criticizing it as exemplifying "the contradictions in the use of creolized practices to promote Indian music", and writing (1997–98, 22–23), that aside from the decidedly amateurish quality of most such performances, questions have arisen as to whether all this activity constitutes local creativity or obsequious imitation of India's commercial entertainment culture. Critics have alleged that rather than promoting grassroots creation, *Mastana Bahar* has in fact served to stifle it with an inundation of filmi pop. Accordingly, local film singers are typically praised not for their originality, but as "the Voice of [film singer Mohammed] Rafi", or as "a true imitator". And while the many teachers and semi-professional performers of film dance choreograph their own routines to songs of their choosing, the style is, with a very few exceptions, derived wholly from the jerky, callisthenic Bombay studio idiom rather than from, for example, the graceful, local chutney style, which evolved organically from Bhojpuri folk dance.

As I have noted elsewhere (S.N. Mohammed 2011), Manuel's condescension to the commercial imported cultural form and his praise of the "organically" evolved cultural form(s) suggest that there is something more "authentic" about the local cultural forms. This distinction falls into the trap of essentialized cultural expressions and fails to acknowledge that neither of these visible manifestations might necessarily bear direct relevance to the lived experiences of the performers or audiences. If each of these forms is taken as equally contrived, and the connections with them equally artificial, then it matters little whether the songs and dances are "traditional" or "commercial". What is more relevant to the present discussion is that this programme, along with its less popular but long-running counterpart *Indian Variety* (aired on Sunday mornings), served to maintain an expression of cultural identity in which the Indo-Trinidadian population (or parts of it) were able to anchor their self-images and find voices, views and portrayals that supported their notions of community.

This is not to say that there was anything "authentic" about the commercial talent show known as *Mastana Bahar*. The particular configurations of cultural expression on the programme were not only (often) random combinations of the received global and the local, but sometimes even farcical representations of the cultural idiosyncrasies of the participants and audience. A popular feature on the programme, for example, included a game called Pick-a-Pan in which contestants could win prizes for answering a question about Indo-Trinidadian culture and then take a chance at choosing from one of three tins ("pans"). The prizes included one tin that contained a "booby prize" that included items considered mildly distasteful, such as a rotting eggplant (*baigan*). Both the questions asked and the booby prizes were often calculated to be self-derisive and poke fun at local Indo-Trinidadian culture or heritage. The programme celebrated the run of its forty-fifth season in 2015.

For several years during the 1980s the lone national television station, TTT, embarked on broadcasts of subtitled Bollywood movies aired on Sunday afternoons. There was backlash to these broadcasts from elements in the society who felt that this was an imposition on their viewing. The criticisms, however, were not levelled at the fact that the films were of foreign origin, but rather more frequently at the fact that many of the films dealt with and displayed domestic violence.

Conclusion

Thus from 1947 to 1993, Indo-Trinidadian audiences in Trinidad were limited to a few hours a week of electronic media programming that featured music and other content that was perceived as being directly relevant to their community or expressive of its cultural heritage. During that time, when representatives of that community complained about being thus marginalized, they often faced the criticism from members of the Afro-Trinidadian majority that Indo-Trinidadian content was on the radio or television all the time. The marginalized status of these media forms would change with the introduction of dedicated Indo-Trinidadian radio that marked the territory of what Reddock (2004, 210) has called "alternative cultural space" in Trinidad and Tobago's electronic media.

CHAPTER 5

103 FM and Indian Radio Stations in Trinidad and Tobago

As the government in Trinidad and Tobago kept broadcasting in check by limiting the number of licences for radio and television, the only outlets for Indo-Trinidadian content were limited slots on the few local radio stations, a regular Saturday-evening television programme and an occasional Sunday-morning slot on the lone national television station. As the government liberalized its media policy, the prospects for a radio station dedicated to Indo-Trinidadian-focused content began to emerge, though its success was by no means guaranteed. Here we examine the emergence of that first station and its implications.

103 FM: The First

The Indian radio phenomenon in Trinidad and Tobago owes its existence in part to a violent attempted *coup d'état* in July of 1990. On the evening of 27 July 1990, an armed Islamic group calling itself the Jamaat al-Muslimeen stormed a session of the national parliament in the capital. They held Prime Minister A.N.R. Robinson and other government officials hostage, bombed the police headquarters and also took over the sole existing television station and one of the few radio stations that existed at the time. In radio and television broadcasts (some of which showed well-known news anchors held at gunpoint) the rebels claimed to be in control of the government and the military. The next few days saw mass media emerging as a key factor in the battle for control as the government blocked the signals being used by the insurgents and issued its own bulletins via television broadcasts from military sites. The insurgents were eventually wrested from the parliament and from the television station, but not before at least twenty-four people had been killed and many others wounded (Searle 1991).

Nationalized and tightly controlled media were a reality after decades of nationalist sentiment in the Caribbean region, where recently independent small states viewed mass media as an important national resource to be used in the interest of social development – as did many other developing nations at the time. Yet government control of media resources by investment in state-controlled media and restrictive licensing and strict controls on private media enterprises, the norm of the time, resulted in few media outlets. During the 1990 coup attempt in Trinidad and Tobago, the vulnerability of the existing national information infrastructure became evident in the ability of the insurgents to capture the sole television station and one of the two radio operations. In control of this major portion of the media outlets, the rebels were able, for a time, to convince the population that they were in charge. Government attempts to transmit competing information reached only a small portion of the population.

The weakness of the information infrastructure under the nationalist/government-controlled paradigm was clearly one of the precipitating factors in the government's liberalization of telecommunications policy to enable the licensing of private radio and television stations, which followed within a few years of 1990. It was not the only factor, since discussions about media liberalization had been in the works for several years before the attempted coup, with parliamentary papers and proposals having been produced in the 1980s and a regional push for liberalization emerging out of regional meetings in 1987.

Among the new television and radio stations was the first radio station dedicated to content aimed at the local Indo-Trinidadian community (commonly referred to as East Indians, to distinguish from the term "West Indian" that is often used to characterize the Caribbean). On 5 July 1993, 103 FM went on the air with Indian and Indo-Trinidadian music, news, announcements and educational material (Mohammed and Thombre 2014).

Something more than a simple radio station or a source for a particular genre of music, the station quickly grew into a nexus of cultural flows in which local artistes could reach a nationwide audience and audiences could interact with musical and media personalities. In a short time, the station spawned competition – some newly minted and others converted into this format – leading to the development of an entire sector of Trinidad radio, with seven stations in 2013 (ibid. 2014) and an additional one having started since then. This expansion of the sector did not go unnoticed internationally. The *Times of India*, for example reported in in 2002 (Viral Bhayani, "Stay Tuned", 2 January 2002, B7) on the proliferation of radio stations in Trinidad specializing in Indian and chutney music.

Among the figures involved in the start of 103 FM is the (perhaps unlikely) veteran radio personality Dik Henderson, whom commentator Keith Subero ("Back to the Sorry Future", *Trinidad Express*, 25 September 2011, 4) described as a former "star broadcaster at Radio Trinidad" back in the 1960s and 1970s. Henderson was later involved in important developments in Caribbean media such as the launch of the Caribbean News Agency's CANA Radio in June of 1984 (Cuthbert 1990), as well as the development of radio much further afield in places such as Tuvalu and Kiribati in the South Pacific with the United Nations Educational, Scientific and Cultural Organization (UNESCO 1983). Henderson eventually became a high-level media manager in Trinidad and Tobago. He was serving as the chief executive officer of the new Caribbean Communications Network group in the early 1990s when an awardee of one of the new licensees named Winfield Aleong approached him (through his friend and co-worker Marcel Mahabir) with a business proposal, as Henderson (interview by author, 15 July 2015) related it:

> Winfield Aleong, through his friendship with Marcel Mahabir, came to us. Marcel was working with me as director of programming. He came to us, brought in by Marcel, to discuss what CCN could do with his licence. . . . So this guy comes along, Winfield Aleong, and he has a proposal, but he doesn't know what to do. He has no background in radio or anything like that. He had background in advertising.

Aleong had been involved with television advertising sales at Trinidad and Tobago Television in the early days of its development (Gordon 2006) and branched out into advertising later. After discussions among Henderson, Mahabir and Aleong, the initial proposal was for a multichannel offering in which Indian music would be one of several components, including sports and other content. However, when they were unable to negotiate this deal with existing media operations, the concept of an all-Indian-content station began to take root. As Henderson (interview, 15 July 2015) recalled, this all-Indian-format proposal also failed to find support with existing media and required Henderson and his compatriots to start from scratch on their own:

> So we set about forming this first-ever Indian radio station in my living room and dining-room table up in Maraval there, in La Seiva. And at that time I was scrambling to get to know a lot more about the Indian community. So in La Seiva they had a lot of Indians there . . . so I was learning as much as I could from them; from Marcel, of course . . . I was then scratching all my contacts in the Indian community to find out as much as I could. Marcel did all the sales and programming, and I did all the business aspects of it, all. And it was quite instructive, an education, to learn how to go about bureaucracy in Trinidad to get things done, having just come back from nearly thirty years overseas.

Difficulties arose because the very concept of a private radio broadcast licence was new to many in the bureaucracy that the founders had to navigate in order to start their operations. Yet as Henderson (ibid.) noted, the difficulties were more than just logistical: "A lot of people I went to . . . in the government, to get things done like VAT [value-added tax] registration and this and that, they were saying to me, like, 'It's treason to have an Indian station . . .' People were really negative about it. But we proceeded and we set it up, and I think the rest is history. It took off."

At its launch in 1993, 103 FM received little mainstream media attention. This lack of attention may have been partly due to the fact that several local corporate conglomerates with media enterprises viewed 103 FM not only as a competitor for market share but also as a misguided upstart for aligning its programming so completely along ethno-cultural lines. The two major daily newspapers at the time, the *Trinidad Express* and the *Trinidad Guardian,* were owned by well-established corporate interests with competing media enterprises, and neither offered substantial coverage of the 103 FM launch on or around its launch date. A full-page advertisement in the *Trinidad Guardian* of 5 July 1993, for example, took this interloper on directly, using graphs to demonstrate that Radio Trinidad (operating on 730 AM since the 1940s) was the main choice of audience members tuning in to radio for both local and foreign Indian music. Perhaps even more pointedly, the wording of the advertisement featured as its headline the question: "What about the other ½ of the population?" with the main body reading as follows (emphases in original):

Radio Trinidad 730 AM is #1 for East Indian programming.
Fact 1:- 47% of this country's population is East Indian.
Fact 2:- No other radio station delivers that target as well as Radio Trinidad.
730 AM IS #1 AGAIN BECAUSE OF YOU!
After all, we have been part of local East Indian Culture for over 46 years with some of the nations [*sic*] most popular and outstanding announcers . . .

The advertisement was replete with spacing and grammatical errors, which might suggest hasty composition. However, it continued to press the advantages of its established radio presence by emphasizing the names of several of its announcers – Farzan Ali, Moean Mohammed and Pat Mathura – whose names would, no doubt, resonate with the audiences for Indian music in Trinidad. Indeed, for several years, the Trinidad Broadcasting Company (TBC) would continue to use its appeal of tradition, as in 1998, when the group wrote to Kamaluddin Mohammed to invite him to host a special programme for Eid-ul-Fitr observances that year – an invitation which Kamaluddin accepted.

This perception of Radio Trinidad's (730 AM) being the original source of Indian music in Trinidad and therefore being important to the Indo-Trinidadian community was not necessarily a new concept or simply reactionary. Evidence of the importance of this content to the station, and of the reciprocity of the relationship between the station and the community over several decades, may be gleaned from a report in *Broadcast* magazine ("Research Figures for Radio Trinidad", 21 May 1979, 19) which noted that the station (then part of the British Rediffusion group) enjoyed high ratings and pointed out that "Trinidad is very much a multi-racial society, with a substantial East Indian minority and programmes of Indian music account for the high ratings as do Indian programmes and transcribed religious programmes".

Despite its claims to being number one and to the tradition of Indian-oriented content in Trinidad and Tobago, 730 AM's limited Indian programming was clearly not sufficient to compete with the full-time dedicated Indian-oriented station in 103 FM. Within just over two years of the launch of 103 FM, the conglomerate-owned TBC responded with the launch of its own Indian station (the fourth station under its control at the time) called Sangeet (a Sanskrit and Hindi word referring to song and dance). Sangeet 106 FM launched on 24 September 1995 and branded itself as a "Superstation", touting both its claims to tradition and support from local Indian music elites. Verma (2000, 356) described the launch of this competitor and another in the following terms:

The great success of 103 FM prompted a number of non-Indian companies to get together and establish their own "Indian" radio station. 106 FM is essentially owned by a group of Syrian Whites, who, in seeing the stunning economic potential resting in the Indo-Trinidadian community, promptly set up their very own "Indian" radio station, replete with Indian songs, talk shows, Indian hosts and programmers, etc. Following suit was 90.5, yet another "Indian" radio station owned by a White group.

The owners of the TBC group, who launched the first competitor, Sangeet 106 FM, followed on their efforts to compete with another such station more than a decade later with their March 2007 launch of Aakash Vani (from Sanskrit, meaning "voice from the sky") on 106.5 FM.

In the face of resistance and direct competition from the established media, which felt that their contributions entitled them to the tradition of Indian radio, 103 FM was still able to stake its claim. While TBC touted the names of some established voices in Indian radio broadcasting (including Kamaluddin Mohammed), 103 FM was able to leverage other established names in Indian music programming from its inception. Indian music programming veteran Hans Hanoomansingh, for example, recounted leaving the National Broadcasting Service and Radio Guardian to join Dik Henderson, Marcel Mahabir (whom he credited with the initial idea for the station) and current manager Hemant Saith for the preparatory work on 103 FM. Hanoomansingh emphasized his ongoing formal and informal relationships with the Indo-Trinidadian community as a key component of his contribution to that initial work. These organizations included the Sanatan Dharma Maha Sabha association (Maha Sabha) – a major Hindu organization that would eventually fight a drawn-out legal battle to establish its own station years later. He also specifically mentioned the Muslim organization (the majority of Trinidad and Tobago's Muslim community is of Indo-Trinidadian descent) known as the Anjuman Sunnat-ul-Jamaat Association as well as the then relatively new National Council of Indian Culture. In his words (interview by author, 15 July 2015): "I was involved as a producer, presenter and president of the National Council of Indian Culture. That is how I was able to invite my fellow directors to meet the Indian community. So I took them to meet the Maha Sabha, and ASJA [Anjuman Sunnat-ul-Jamaat Association], and other organizations, and I myself represented the National Council of Indian Culture."

While the traditional sources of limited Indian music programming tried to protect their traditional niches and their dominant mind-set about the place of

Indian music in the national mediascape, something decidedly untraditional was happening at 103 FM. The station was redefining the broadly accepted notion in Trinidad and Tobago that Indian cultural expression was appropriately ghettoized outside of mainstream programming. As noted above, radio time for Indian or Indo-Trinidadian music included fairly limited slots such as fifteen-minute segments on each of two stations on weekday evenings and one-hour segments on Sunday mornings. This was expanded to include one-hour weeknight segments on one station (the state-owned National Broadcasting Service, on 610 AM) during the late 1980s and early 1990s. Hanoomansingh (ibid.) argued that the introduction of 103 FM marked an innovation not just in the media history of Trinidad and Tobago, but also in the identity politics of the Indo-Trinidadian community: "From a broadcasting point of view, it gave the community greater exposure, more broadcasting time. From a cultural point of view, from a sociological point of view – I think that this is the most critical thing – it's provided the space and identity for the Indian community that was missing. We were limited to fifteen minutes and half an hour. Now we have this station of our own, the community's station." Hanoomansingh also made the connection between the emergence of 103 FM and the establishment of the Divali Nagar – a major cultural event held annually that celebrates not just the Hindu festival of lights, but also Indo-Trinidadian (particularly Hindu) heritage more generally.

Several of the figures involved in the early days of 103 FM recalled that the experience was something more than simply the launch of a business, describing it as emotional for both themselves and the listening community (Saith, interview by author, 15 July 2015). Recounting the emotion of the early days, Dik Henderson (who, as noted above, is not an ethnic member of the Indo-Trinidadian community) also noted that "people would come in with bags of money to give us to make sure we survived. . . they wanted to pay for requests, all kinds of things" (ibid.). Hanoomansingh (interview, 15 July 2015) recounted a similar experience in somewhat greater detail and connected the gesture of monetary donations to the identity function of this first Indian music station:

> I remember a man from Rousillac who came, . . . He arrived at about 6 o'clock in the evening. It was probably the third or fourth day we were on the air. He had in his pocket $600 and he said, "I have come to buy some advertising."

> He didn't come to buy advertising: he had a scrapyard. He had never advertised before. He was so overwhelmed. He wanted to, in a tangible way, express his feelings. And the only way he could do it is to come and say, "I want to advertise and I have $600." But that was the feeling that came out of identity: "This is our station."

Henderson (ibid.) corroborated this characterization of the response to 103 FM, indicating that many listeners brought their record and cassette collections of Indian music to donate to the station and to support its activities. Describing the response to 103 FM as a "groundswell", Henderson (interview, 15 July 2015) noted, however, that the early operations were limited to a few hours a day and even those few hours posed challenges, due to staffing and other limitations: "We had a very small staff. Hemant [Saith] was doing several different things at the same time. I was writing commercials, plus at one point I was on the radio between twelve and one because we couldn't find an announcer to do it. So I had an operator just spinning the music with me and I was giving the time – not pronouncing the Hindi words at all." With rented computers, a small office with a single studio and a skeleton staff, the station at first broadcast content for a few hours a day and had to cease transmission at 8:00 p.m. to allow for production of materials for the following day's programming (ibid.; Saith, interview, 15 July 2015). According to Henderson, the team faced a tremendous amount of scepticism from others within the radio industry and its advertising base; much of the limited support from advertisers in the beginning was leveraged, in part, on the reputation that Marcel Mahabir had built over many years in radio advertising sales (and even those bookings were tentative at first).

Verma (2000, 355–56) described the new Indian stations, particularly 103 FM, as bringing greater recognition, visibility and status to the Indo-Trinidadian community that quickly translated to economic influence and advertising revenue:

> Given its massive listenership, a lot of people began paying more attention to the Indian community. For instance, in the business world, the commercial world, the music and religious world, the Indo-Trinidadian came to be recognized as a highly attractive and valuable consumer. Local companies in Trinidad were beginning to realize the strong purchasing capacity of the Indo-Trinidadians, and started advertising campaigns specifically targeting the Indo-Trinidadians.

Hanoomansingh (interview, 15 July 2015) noted that the growth in advertiser buy-in was initially slow but accelerated as demand for the station's content and grassroots support quickly drove 103 FM to expand its broadcast offerings and address a national audience:

> The initial support for 103 came from the Indian business community, not the advertising agencies. [After] the initial surveys and the public response, when the advertisers saw the overwhelming reaction, they started supporting. But initially that support was not there. The encouragement came from the Indian businessmen, from the Indian community. So we had, in a matter of weeks, to go to twenty-four hours, and we had to change the antenna in order to reach the entire country.

This support was not limited to the business community, as there was evidence that the political leadership had taken note of the station and its impact as well. Hanoomansingh (ibid.) recalled meeting the leader of the opposition at the time (who would later become prime minister), Basdeo Panday, at an event and being told of the impact that 103 FM was having:

> I went to announce my arrival to my hosts and then I felt a hand on my back. And I looked around and it was Basdeo Panday. He said: "I told my friend . . . I must come and talk to you, and congratulate you for 103 FM." He said: "You know, I'm a politician, I'm out late in the night, and there's no greater feeling than driving home from a political meeting and as I get into my car, I put on the radio, and I'm hearing Indian music." And in the context of Basdeo Panday, it was more than hearing Indian music, it was a big statement, it was a big occasion.

The launch of 103 FM and the evolution of similar and competing stations in later years, along with the related developments in Indo-Trinidadian cultural expression, including chutney and chutney soca, have sometimes been cited among the factors that saw the ascension to national leadership (for the first time) of political parties with predominantly Indo-Trinidadian membership (see, for example, Verma 2000).

Yet, even while the station became popular, advertisers flocked to buy time and its programming developed into a full twenty-four-hour offering, the vision of its content was by no means a settled matter. Henderson described a programming strategy that depended on the musical knowledge of the announcers (including some who were related to Kamaluddin Mohammed), who had some experience with the genre and some understanding of the audience. However, this was an untested audience in the context of full-time broadcasting, where the demands of a full programming day demanded a larger set of music choices that the traditional fifteen- or thirty-minute slot to which many of the deejays were accustomed and a departure from the steady diet of orthodox Hindi film music (the trade in which had locally been largely dominated by a few people – also including relatives of Kamaluddin Mohammed). This meant that a greater variety of music was not only possible, but in some ways, necessary. Among the first challenges that the station faced regarding its programming was controversy over playing chutney music. Henderson (interview, 15 July 2015) described the problem in the following terms:

> We had a mix of – this is quite ironic – we had a mix of Bollywood music and chutney. And I'll tell you, we had a petition within the first week of playing chutney on the radio. And all this is happening within the first month, or six weeks. We had a petition – that's thousands of people objecting to us playing chutney. And I tell you that – I don't know what became of these things later – but we had a paper we could have rolled out all down the front, all down the stairs into the car park.

Radio 103 FM's initial experimentation with content and format included not only controversy over its use of chutney but also some other bizarre twists such as the fact that, for some time, the station played country and western music because research showed that this genre was popular among the local Indo-Trinidadian population. Chutney continued to compete with foreign Hindi film music on 103 FM while also gaining in popularity in the broader society, though its place on 103 FM continued to be a source of contention. According to Henderson (interview, 15 July 2015) and Saith (interview, 15 July 2015), their accession to chutney led to concerns about this genre becoming dominant and the consequent loss of what they characterized as "middle-class" audience members. Their eventual strategy with this genre has been to relegate chutney to the weekends.

Yaseen Rahaman (interview by author, 15 July 2015) placed the station's programming challenges in the context of an evolving market in which Trinidad's Indian-format radio has developed into a much more "niche-oriented" market, with several segments that have evolved as important, including such designations as youth, religious, mature, Bollywood and others. These developments have required strategic repositioning and adaptation to deal with increasing fragmentation.

Alongside its struggles to develop programming that its audiences found attractive, 103 FM soon also encountered the establishment of other Indian-format stations that competed for audience share. Despite the challenges of an evolving market and an evolving audience, 103 FM management is confident that they remain ahead of the game and continue to act as the leaders in their market segment. Henderson (interview, 15 July 2015) argued that the station has been able to capitalize on its pioneering role by establishing loyalty from its listenership base:

> The core listenership remains, up to this day, as people who are loyal to 103 FM and the first Indian station in Trinidad. . . . People in, by and large (and particularly the older) Indian community, are loyal to us. . . . The other stations, having come on two years later, they're always playing catch-up. What also distinguishes us is that we shifted our focus under Mr Rahaman's stewardship to public events involving people, outreach programmes – going out there. And so we have a lot, a number of public events – which unfortunately our competitors copy. So we have really down-to-earth things like the "curry duck", the *mela*, the road show, "Christmas bandwagon". . . . We go out to the public, and . . . in the last fifteen years, sixteen years these things have grown . . .

Competition, Imitation and Conversions

When it became clear to the local market that 103 FM had quickly become a commercial and cultural success, several competitors from the established mainstream media began to create offshoot Indian radio operations. Despite the emotion and identity claims surrounding 103 FM, Hanoomansingh (interview, 15 July 2015) argued that the station was primarily a business enterprise and, due to its success, spawned a number of imitators:

> 103 stumbled on an idea that was instantaneously successful. 103 did not recognize the dimensions of that success, the yearning of that community. The rule of 103, in terms of space and identity, it was a business venture. It didn't have the objectives of, say, the National Council of Indian Culture, it was a platform for a business model that became very successful. . . . So it was purely entertainment and return on investment. . . .
>
> I remember at a board meeting Dik Henderson saying, "You know, somebody, some entity, will start another station playing Indian music." He was absolutely correct: 106 came shortly afterwards, then 90.5 and so on.

For Hanoomansingh (ibid.), the proliferation of Indian-format music stations in Trinidad and Tobago comes with some failings, particularly in their propensity to cater to a limited set of musical expressions that ignore pioneers of Indian and local Indian music – forms that were more likely to be heard on the limited-time Indian programming in years past. Hanoomansingh argued:

> If you take all the stations together, with the emphasis on the business model and business success, in a sense, with more eyes dedicated to Indian music, we've also lost some of the element that prevailed when we had limited programming. So there's a question as to whether more music, more Indian music, more stations have contributed to the development of the culture. And some areas have suffered, like, for example, the classical Indian music.

At the time of writing, there were some nine stations that may be categorized as Indian-format broadcasting in Trinidad and Tobago. Mahabir (2016, 2) identified the nine stations falling into the Indian-format category as comprising the pioneering 103.1 FM, Hanoomansingh's Heritage Radio on 101.7 FM, two stations owned by a local media conglomerate (Sangeet, operating on 106.1 FM, and Aakash Vani 106.5), Radio 90.5 FM (which converted from mainstream easy-listening to Indian format in 1997), the Hindu-focused Radio Jaagriti 102.7 FM (covered in greater detail below), Couva (in central Trinidad)-based station Radio U97.5 (which styles itself "Hot like pepper"), and the newest addition, Taj 92.3 FM, launched in 2014 as part of a larger broadcasting operation featuring a variety of stations. WIN Radio 101.1 FM is included in this count and its hybrid approach is examined in detail in a later chapter. However, having inherited the operations of the former Masala 101 FM, the station ceased to broadcast during 2015 after failing to meet licensing-fee and other financial obligations.

While some of these operations have developed a constant presence on the air and most have migrated also to including online echoes of their broadcasts, their market niche has also seen stations come and go in a variety of business scenarios. WIN 101 FM, for example, emerged out of the collapse of a station known as Masala 101 after the earlier operator on the frequency went out of business, as its parent company (the Hindu Credit Union [HCU]) crashed. Another station associated with the same crash, Radio Shakti, which operated as a contractor to the HCU, converted to an independent operation on the same frequency and operates as Hot Like Pepper Radio on 97.5 FM.

All this reflects some level of volatility in the Indian-format radio market niche in Trinidad, driven in part by the attraction of great enthusiasm in its audiences, but limited by its small size and its fragmentation.

Conclusion

While the Indian-format subsector of Trinidad's radio industry is today vibrant and competitive, what emerges from the words of some of its pioneers is a picture of a venture that faced both derision and opposition at first. 103 FM, faced with the challenges of capital and forging a market that was unproven and uncertain, also found itself with waves of emotional and financial support from its target audience, members of which may have imagined such a station only as the most remote of possibilities. Established elements of the local media which felt some ownership of Indo-Trinidadian programming and whose reputations had been built on providing such programming on local media responded at first with resistance, but before long, such resistance was replaced with competitive efforts aimed at tapping the new and previously unrealized market.

Within the complex cultural matrix of Trinidad and Tobago, the establishment of 103 FM and its later competitors was not without its impact on identity and ethnopolitics in the small nation, coming as it did in close temporal proximity to other developments in which Indo-Trinidadians were experiencing some levels of social and political advancement. As Fergusson (1997, 14) noted: "In the wake of the Indians' political power, a proliferation of Indian radio stations, television programming, and other Indian cultural phenomena has added fuel to Afro-Trinidadians' fears in the rapidly changing bureaucratic and economic landscape, creating frustration among the large number of disenfranchised blacks." Thus even in its unsure attempts to establish a cultural voice for a culturally disenfranchised group, 103 FM became part of a cavalcade of events that may have marked it as a source (or at least a component) of disenfranchisement.

CHAPTER 6

Media, Religions and Radio Jaagriti

The evolution of the Indo-Trinidadian-format radio subsector in Trinidad and Tobago has included diversification into even more specialized broadcasting operations. A station dedicated to the local dominant understanding of Hindu religion called Radio Jaagriti has emerged as one such specialist operation. Eschewing much of the raucous content and party promotions of its competitors, this station engages exclusively in religious instruction and entertainment. Yet, as the local radio environment grew and numerous other entities were being granted licences to become radio broadcasters, the parent body of Radio Jaagriti perceived that reluctance to grant approval for its application was motivated by racial and/or political prejudice, a matter which resulted in a prolonged legal battle for the station's existence, which we examine in the present chapter, along with the broader context of religious content on Trinidad's Indian-format radio.

Religion and Indian Radio

The religious component of Indian radio had been evident since the start of radio in Trinidad. Kamaluddin Mohammed's inclusion in the opening ceremonies of Radio Trinidad and his ability to convince the station's administration that he could represent the Indian presence in Trinidad on the radio was partly a result of his ability to read, write and speak several pertinent languages but also, arguably, partly due to his recognized accomplishments as a religious scholar. The argument weakens with the fact that he was primarily a scholar of Islamic religion, which was a minority religion among the indentured labourers and today accounts for around 10 per cent of the population of Trinidad and Tobago. Yet the multi-religious ceremonies at the start of Radio Trinidad and the persistence of religious programming on radio demonstrate the importance of religious content as part of both Trinidad radio in general and Indian radio in Trinidad in particular.

For an informed perspective on the role of religion in Indian radio in Trinidad, one of the people the author interviewed during the fieldwork was Pundit Jeewan Maharaj, who has served as a religious contributor on several radio stations over the years. The term "pundit" designates a religious leader or priest in the Trinidad Hindu community. Maharaj is also trained professionally as a lawyer.

Prior to the establishment of a dedicated Hindu station, most of the Indian-format stations featured some form of religious programming. These were predominantly Hindu programmes with some Islamic content, particularly around Ramadan and Eid-ul-Fitr. One early addition to the religious offerings, for example, was the inclusion of Hindu astrological readings in place of traditional Western astrology that could still be heard on many mainstream stations.

When 103 FM started operations, Hans Hanoomansingh invited Pundit Maharaj to participate in a morning devotional programme called *Thought for Today*. This early-morning programme proved something of a challenge, as the pundit did not know if anyone was listening at that early hour – though listener feedback he received later on surprised him as to the extent of his audience. The programme included a small prayer segment and some commentary on religious and philosophical matters. Pundit Maharaj indicated that, despite being a Hindu religious figure on an Indian radio station, he still felt the need to deliver messages that had what he characterized as a nation-building quality. Eventually he would become a popular draw for several local media outlets, but says that his longest stint was his time at the Indian station now known as 90.5 – the People's Station, emphasizing its cosmopolitan approach:

> The other stations would have me as a guest, to interview or to speak, but the longest stint was at 90.5, which I enjoyed tremendously – a lot of talented people there, they knew the music. They would set up the music and so, which was good for me because

Christmas time, you know, they would put on appropriate tunes, and Eid and so on. At fasting time we had some nice . . . songs appropriate to that, Divali times. And then they'll mix them generally.

So here I am in a cosmopolitan country, I didn't take on as much at first, but the people on the outside were listening, which I realized later through e-mails and so. But I tried to give a nation-building message.

Pundit Maharaj viewed his religious mission on Trinidad's Indian radio stations as not so much one that solely promoted his own religious teachings, but a mission to expose citizens to the multiplicity of people and views that inhabit the country's social landscape. He noted in the interview that he had the opportunity to present a range of ideas and cultural differences to his audiences and the opportunity to get their feedback as well, saying:

If somebody doesn't like you, sir, it's a good chance they don't know you, that's why. . . . So give them a chance to see that maybe a Trinidadian and Tobagonian could be a nice person. Let them see that an Indian could be a good person. Let them see a Hindu could be a good person. There are good Hindus; there are good Muslims; there are good Christians. . . . What helped me too, were the responses that we got – who was calling in. A Hindu would call, a Muslim would call, a Christian would call. And then you're like, "Oh, here are the kinds of people calling, the kind of people I'm bringing my message to."

These were not only people responding to religious messages or deep inspiration. Truck drivers would call in, saying that the pundit's breathing exercises were helping them to relax on the road. His callers included people from the Afro-Trinidadian community who may have come across his messages on the radio by accident, but who were moved to express their appreciation. Messages from diverse listeners and from listeners far away, eventually even those in foreign countries responding via calls and e-mails, convinced Pundit Maharaj that his messages of tolerance were being well received. He noted, however, that he would always avoid what many call comparative religious commentary, explaining that once we compare, someone wants to come out better than the other person. Pundit Maharaj noted that this new dedicated Hindu focus of Radio Jaagriti and its significance and ramifications needed to be examined.

On the current state of Indian radio generally, Pundit Maharaj expressed concerns about the commercial appeal of religious debate, some of which can be divisive and acrimonious:

I'll put this part bluntly . . . If you have a business, the purpose of your business is that it tries to make money – how do I make money? Even if I have a religious programme, am I screening the people who are talking? Suppose the media comes and pits two religions against each other? The point is, we're getting a good advertisement, we're getting sponsorship, you could say what you want.

Is there a screening process for the people who are touting religions, who are speaking about religion? Let's say I have a station and some fellow comes in and says, "Only a Hindu could be your friend, only a Muslim could be your friend." That's not how we grow up and pitching marbles and playing cricket in this country.

Pundit Maharaj did not, however, place all the blame on the media. He saw an inherent unity among the peoples of Trinidad and Tobago on an everyday basis, but argued that the very religious institutions that should be promoting tolerance instead emphasize difference and discord: "It's when you went to your place of worship they started to teach you about hate and difference. On your own you're getting along well. . . . But the fellow who you depended on to teach you peace, and togetherness, and brotherhood and love – he was the one saying he's different. "We're good; they're bad."

This kind of discord makes for good programming, and the radio stations can capitalize on the popularity of divisive religious ideas in many ways. For Pundit Maharaj, the motivations for religious programming can be somewhat crass:

So therefore, the people in charge of the stations . . . they're thinking, "Well, I have a business here, how can I compete with him to make money? Ok, the people want to hear a religious programme. I don't really care for a religious programme, but the religious presenters will pull in a lot of listeners, so we can get more advertisers and we can have a bigger salary and get more money."

For Pundit Maharaj, these crass motivations for religious content can lead to curious double standards in programming, particularly on Indian radio stations, where advertisements for alcohol and parties can coexist with religious programming at times when abstinence is required by religious adherents (as mentioned elsewhere, Radio Jaagriti maintains a strict sanction against advertising alcohol). However, his primary focus was on what he perceived as the opportunities for national unity and harmony afforded by the multiple currents of religion in Trinidad and Tobago and their presence on the airwaves: "We have a unique lesson in Trinidad and Tobago that could put all of us of different cultures here together, and we have a chance to share and develop and grow because of it. If I were a Hindu alone here, for example . . . I would be benefiting from what I could learn from Christianity. I could benefit from the pillars of Islam."

The Battle for Radio Jaagriti

Despite a plurality of religious expressions on the radio and broader media in Trinidad and Tobago, the experience of Trinidad's lone Indian-format station focused on the Hindu religion has been something less than an exposition of pluralism. This particular station, in fact, emerged only after an acrimonious and hard-fought legal and social battle for its existence. By the time Radio Jaagriti ("Reawakening") was launched in 2007, it had already become a major story in the local and at times even the international press. Coverage of the not-yet-existent station had less to do with its specialist mission of Hindu religious broadcasting than with a prolonged legal and social debate over the granting of a licence for such a station. That battle lasted seven years and involved local courts, a suit against the Attorney General and appeals all the way to the highest legal authority recognized by Trinidad and Tobago at the time – the British Privy Council.

The organization behind the proposed Radio Jaagriti was the Sanatan Dharma Maha Sabha (SDMS/Maha Sabha). Formed in 1952, this organization would evolve into the most influential and prominent Hindu organization in Trinidad and Tobago (Malik 1966; Vertovec 1994). It emerged as part of a movement in the 1930s among Hindus in the British colonies of the West Indies to organize and consolidate the religious ideas of the migrants. In the early decades of the twentieth century, religious debates raged in India between reformist groups such as the Arya Samaji and so-called Sanatists, or Brahmin-based religious groups holding to the traditional caste-based authority of the priestly sect. These debates spread to the Caribbean, where missionary representatives of each strain of Hindu thought courted the allegiance of the displaced indentured populations. According to Vertovec (1994, 138) the "Sanatanists" in the region succeeded to various degrees in achieving reforms from colonial authorities such as recognition of Hindu marriages and the right to perform cremations, but it was Trinidad's Sanatan Dharma Maha Sabha that was the most successful, with its founder, Bhadase Sagan Maraj, uniting rival organizations (the Sanathan Dharma Association and the Sanatan Dharma Board of Control) to form the Maha Sabha in 1952. Malik (1966, 91) described the founding of the Maha Sabha in the context of Hindu revivalism in the 1950s, noting the role of caste (through generally already quite deprecated as a social force in the Indo-Trinidadian community by that time) and politics as follows: "In this Hindu revival the *Sanatan Dharma Maha Sabha* and its high caste leadership played a very significant role. . . . Under the Presidency of Bhadase Sagan Maraj, the Sanatan Dharma Maha Sabha became one of the most powerful religious and political forces among the Hindus of Trinidad. It has become the strongest base of the political strength of the Hindu community."

Malik's recounting of the importance of the Maha Sabha as a political force ("an important agency of particularistic political socialization of Indo-Trinidadians" [ibid., 94]) in the 1950s also exposed something of the caste dimension underlying this power and notions of the legitimacy of power being based, at least in part, on the credentials of blood and heritage:

> A candidate seeking election in the Hindu majority areas of Trinidad would be running the risk of losing by openly criticizing the *Sanatan Dharma Maha Sabha*. In the 1956 elections L.F. Seukeran, who was seeking election as an independent candidate was charged by his opponents with being against the *Maha Sabha*. Seukeran promptly denied the charge and said, "I am a Brahmin and son of a Brahmin. Seven generations of Brahmin blood flows in my veins. I know more Sanskrit and Hindi than all the Pandits in Debe." (Ibid., 91)

MacDonald (1986, 107) wrote of the Maha Sabha that it "aimed at reviving and maintaining the Hindu religion and culture from the inroads of Christian missionary activities", becoming "a powerful political force during the 1950s" in predominantly Indo-Trinidadian areas. MacDonald (ibid.) also recounted the emergence of Bhadase Sagan Maraj, who was born in 1919 in the East Indian enclave of Caroni, as a combination of political and economic factors:

> Maraj was the son of the village headman, a traditional authoritarian figure who was a self-appointed judge, solving all disputes. At an early age Bhadase became involved in politics since his father was shot and killed at home and his brother-in-law was strangled by another village faction. Saving his money from work in the cane fields, he soon bought a boat and established himself as a small-scale contractor for construction on the U.S. bases. At the close of World War II, Maraj bought a considerable amount of equipment from US personnel, which made him somewhat

wealthy. Financially established, he turned to politics and was elected in 1950 as an Independent on the Legislative Council.

Maraj figured in the politics of Trinidad and Tobago not only as leader of the Maha Sabha but also as founder and leader of a political party known as the People's Democratic Party (PDP), which formed the opposition in Trinidad's legislative council prior to independence. According to MacDonald (ibid.), Bhadase Sagan Maraj's PDP emerged in part owing to perceptions of racial undertones to Eric Williams's PNM, which was perceived as excluding the Hindu community, which formed the majority of the Indo-Trinidadian population. MacDonald (ibid., 107) wrote:

> Although the PNM had repeatedly emphasized that it was a multiracial organization, it was largely Afro-Creole with small numbers of Christian and Muslim East Indians and a few white Creoles. To the majority of the Hindu East Indians, the PNM represented nationalism, but Afro-Trinidadian nationalism . . . In response, the Hindu community turned to the People's Democratic Party (PDP), which soon became the second largest party in the colony.

Additionally, for MacDonald (ibid.), Maraj's role in the Maha Sabha bolstered his national political aspirations and "since Maraj was the president of this religious and cultural organization, it was only natural that the SDMS strongly supported the PDP, which was founded the following year by the same individual".

Bhadase Sagan Maraj's power was also consolidated through his involvement with the sugar workers' trade-union movement, since Indo-Trinidadians were the primary labour component of the sugar industry at that time. Knowles (1959, 84–85) wrote of this influence and how the combination of religion, politics and power were mixed with negative perceptions of Maraj:

> In 1953, the All Trinidad Sugar Workers Union and the Sugar Industry Workers Union agreed to federate under the presidency of Bhadase Maraj, a businessman who is alleged to have made a fortune under questionable circumstances dealing in United States war-surplus goods. He formed a quasireligious political party and was elected to the legislature in 1952. . . . Prominent Trinidadians interviewed by the writer went further to accuse Maraj of bringing Chicago-style racketeering methods to Trinidad and feared that he would succeed in joining together in [an] Al Capone manner politics, business, trade unionism, and religion.

To make the perceptions even worse, Figueira (2003) noted that the belief was widespread that Maraj, while being the head of the sugar workers' union, was also being paid by the colonial owner of the sugar factories and plantations, Tate and Lyle, to keep the workers docile.

The formation of the SDMS and its increasing visibility, along with its appeals to Hindu and Indian identity, coming in the wake of the independence movement in India, led to political backlash from Eric Williams and the PNM. Sherry-Ann Singh (2005, 357) has noted that Williams portrayed the PDP as "a reactionary, communal Hindu organization" and consistently attacked the SDMS with "general anti-Hindu tactics", noting that "while the formal policies of the PNM would point towards a vision of inclusion of 'every creed and race' and the promotion of secular values, the actual implementation, far from a realization of the 'objectives', served to heighten the level of alienation felt by Hindus in almost all spheres".

The global scope of influences and discourse of the time, already aided by radio and other flows of international news, was such that, leading up to the 1956 electoral campaign, Eric Williams was able to invoke fears of violence from the SDMS (sometimes simply referred to as the Maha Sabha) by positing a connection between the Trinidadian organization and a similarly named group in India, the nationalist Hindu Maha Sabha, that was implicated in the 1948 assassination of Mahatma Gandhi. As MacDonald (1986, 108) described the issue: "Eric Williams and the PNM linked the SDMS to the similarly named *Hindu Maha Sabha* of India, which had been responsible for Gandhi's assassination. Although the SDMS denied any linkages, pointing out that their's was not a political organization and had been incorporated several years after the Mahatma's demise, its political influence had been made public."

Outside of Williams' ill-founded allegations, other serious mutual misapprehensions would emerge surrounding not only the mobilization of the Indo-Trinidadians under Bhadase Sagan Maraj, but also Williams' PNM and its stance on education at the time. Bhadase Sagan Maraj gained the allegiance of Hindu groups and congregations island-wide and the SDMS oversaw the construction of Hindu schools throughout Trinidad. As MacDonald (ibid.) described the situation, the Catholic Church had earlier established its own group of schools with their own religious and cultural emphases, but the new SDMS Hindu schools catered almost exclusively to Hindu Indo-Trinidadians, with religious and cultural education that was perceived as running counter to the dominant nationalisms:

The fact that few non-Hindu and non-East Indian students attended these schools guaranteed an exclusive Indian cultural and religious education, seconding [*sic*] the Hellenic-Christian-dominated West Indian culture and sense of nationalism favoured by the Afro-Creole community. For the Hindus the school system provided employment for young Hindu teachers and functioned as a bulwark against Christian conversion. For the PNM it was a dangerous factor in continuing and exacerbating the divisions in Trinidad and Tobago's society.

Perhaps to confirm MacDonald's suspicions about the SDMS schools and competing nationalisms, several older collaborators confirmed that it was common practice in the 1950s to sing the Indian national anthem at the start of the day at those schools. However, despite this international observance, one of the SDMS schools drew international attention for a different reason when it was set on fire a few days before the legislative council election in 1956, with Maraj claiming that it was an act of political violence (Paul P. Kennedy, "Trinidad Is Tense on Eve of Voting: East Indians Seek to Wrest Political Control from the Negro Majority", *New York Times*, 23 September 1956, 17).

As the islands and territories then known as the British West Indies debated the British proposal to form a post-colonial West Indian Federation, the *New York Times* (Sam Pope Brewer, "Race Issue Curbs West Indian Ties: Parley of the British Colonies in Caribbean to Study Factor Hampering Federation", 11 February 1955, 6) identified Indo-Trinidadians as a barrier to broader West Indian (Caribbean) integration, because they had a propensity to "resist assimilation into the general population and tend to maintain their ethnic identity", specifically noting the roles of both Maraj and the SDMS in the negotiations toward federation.

When the British colonial authorities' Capital Site Mission issued a report critical of the choice of Trinidad as the location of the capital of the federation owing to the presence of the Indo-Trinidadian community and the presumed social uncertainties that might arise, Trinidad's premier and head of the legislative council – Eric Williams – turned to Maraj for support. Ryan (1999, 168) wrote that Williams "resented the Mission's slur on political life in Trinidad and Tobago and on the Indian community", and noted that in 1957 Williams even persuaded Bhadase Sagan Maraj to travel with him to Jamaica to make the case for situating the federal capital in Trinidad.

The SDMS continued to play an important part in the Hindu community, national discourse and national politics after Trinidad and Tobago gained independence in 1962. Reporting the findings of a 1965 survey of Indo-Trinidadian elites in Trinidad, Malik (1970, 555) noted that among Indo-Trinidadian politicians and trade-union leaders who demonstrated a religious affiliation, the SDMS was a dominant force, writing:

> The *Maha Sabha* leadership is in the hands of high caste Hindus, as is the political leadership of the East Indian community. Since the bulk of the Hindu population belongs to the Sanatan Dharma, political leaders of the East Indian community have skilfully used the *Sanatan Dharma Maha Sabha* for consolidation of the Hindu rank and file. . . . The Maha Sabha dominates the business elite too; 38 percent of them are members of that organization.

Bhadase Sagan Maraj died in October 1971 and his son-in-law Satnarayan (Sat) Maharaj took over leadership of the SDMS. It was under the stewardship of Sat Maharaj that the idea of a Hindu radio station would emerge and the fight for Radio Jaagriti would be taken to the courts.

The emergence of Radio Jaagriti came only after a legal battle that lasted more than seven years. Public sentiment regarding the case was (at least in part) conditioned by attitudes to the strongly polarizing leader of the SDMS, Sat Maharaj. The reputation of this outspoken community leader and activist spreads beyond the shores of Trinidad and Tobago. Writing in an article for the Associated Press that was widely carried in US and international newspapers, Tim McDonald ("Hindus and Christians Compete for Followers in Trinidad", Associated Press International, 31 May 2000) described Maharaj's response to evangelical preachers and missionaries in Trinidad: "'I told our people to throw these people out of the villages,' said Sat Maharaj, head of the Hindu organization Sanatan Dharma Maha Sabha. 'We launched a counter-campaign' that includes literature pointing out alleged inconsistencies in the Bible and what Maharaj calls its undue focus on material possessions."

For Sat Maharaj, neither the mainstream mass media nor the wholly dedicated Indian-format radio stations were catering to the needs of the Hindu community. In an opinion piece for the *Trinidad Guardian* newspaper prior to the initial launch date of Jaagriti, he wrote ("A New Awakening", 7 December 2006, 5): "The 2000 Report of the Central Statistical Office of T&T showed that the Hindu population in the nation comprised 23.4 per cent. Yet despite our numerical size the Hindu population's religious needs were virtually ignored by the national media and even by the Indian-formatted media houses."

Radio Jaagriti's lengthy legal battle for its existence began in December 1999, when the SDMS applied to the local authorities for a radio broadcasting licence. Two and a half years later, while still waiting for it to be approved, the SDMS learned that a radio broadcasting licence had been awarded to a company called Citadel Limited. Citadel's directors included prominent supporters and financiers of the PNM government, with one director having served as a political operative for high-profile PNM candidates in past elections and later having held key appointments in state enterprises.

In response to the fact that Citadel had been awarded a licence and the SDMS had not, the SDMS and its incorporated company Central Broadcasting Services Limited brought a constitutional motion on 16 August 2002. With the award of the Citadel licence and the failure of the government to grant the SDMS licence, the SDMS claimed that its constitutional guarantee of equal treatment under the law had been violated. It also claimed breaches of its rights to freedom of conscience and religious beliefs and freedom of thought and expression. In the motion it sought to have the courts force the government to issue the broadcast licence.

On 4 February 2004 the local High Court found for the SDMS, agreeing that the organization had been treated unequally. However, the court declined to consider the further claims of breaches of freedom of conscience, thought and expression. More to the heart of the case brought by the SDMS, the court would not, at that stage, issue an order instructing the government either to consider or grant the licence sought by the plaintiffs, arguing that to do so would interfere with the power of the government – the legitimate authority by which such licences were granted.

A week later, on 11 February 2004, Maharaj and the SDMS wrote to the Prime Minister, Patrick Manning, demanding that its broadcast licence should be granted by a deadline of 20 February 2004. Instead of acceding either to the court's decision or the Maha Sabha's demands, the government appealed the decision, challenging the discrimination finding. In January 2005, the Appeal Court threw out the appeal and ordered the government to reconsider the licence application ("Maha Sabha Bid for Radio Licence Turned Down Again", *Trinidad and Tobago Guardian*, 21 May 2005, 9) and issued a second ruling on 21 April 2005 giving the Cabinet twenty-eight days to make a decision on the licence (Imran Ali, "Court: Grant Radio Licence to Group", *Trinidad and Tobago Guardian*, 22 April 2005, 3). However, the government attempted to sidestep the order, claiming that while the court ordered the Cabinet to consider the licence, the granting of licences had been shifted to the new Telecommunications Authority (ibid.).

Despite these manoeuvrings, the courts ruled that the Cabinet still had the power to consider the licence application and the government did eventually deliberate on the application. On 17 May 2005, the Cabinet denied the broadcast licence. Minister of Public Administration Lenny Saith wrote to the Maha Sabha saying that the application had been rejected because the government felt it needed further information that had not been supplied. Sat Maharaj called this decision "the best evidence of discrimination" and claimed that it "showed that Government was in fact biased against the Maha Sabha and intent on continuing its discrimination against the Hindu community" (*Trinidad and Tobago Guardian* 2005, 9).

A few days later the Maha Sabha announced that it had gained permission to take the matter of its licence application and denial to the London Privy Council. On 23 May 2005, the Court of Appeal granted the Maha Sabha leave to take its fight to the British Law Lords ("Maha Sabha Takes Fight to Privy Council", *Trinidad and Tobago Guardian*, 24 May 2005, 8). In July of 2006 the Privy Council ruled for the Maha Sabha ("Privy Council Rules against Government: Maha Sabha Wins Radio Licence Case", *Newsday*, 5 July 2006) and, subsequently, after a few false starts, Jaagriti went on the air on 19 January 2007 at 5:05 p.m. at Pasea Road in Tunapuna ("Maha Sabha's Radio Jaagriti on Air", *Newsday*, 20 January 2007) with a staff of fifteen, some of whom were recruited from existing Indian-format radio stations ("Radio Jaagriti to Begin Testing Dec 1", *Newsday*, 17 November 2006, 3).

With its stated emphases on Hindu religious teachings, ceremonies and celebrations, Jaagriti would not remain immune from the influences of race and politics in Trinidad and Tobago. When the offices of Radio Jaagriti were burgled and vandalized in 2013, Sat Maharaj invoked political interference, saying to a local news outlet that he felt that Jaagriti's pro-government stance made it a target ("Radio Jagriti Vandalized", C News, Caribbean New Media Group, 10 April 2013). While there was no evidence of political motivation, local media noted that suspects in the crime were apprehended from a predominantly Afro-Trinidadian community.

Jaagriti expanded its operations to include television programming, with a cable television operation and eventually a broadcast presence as well. The station also currently streams its content on the Internet. While the Indian music and entertainment radio format is itself something of a niche market operation, Radio Jaagriti finds itself in an even more restrictive situation. Not only does the station eschew non-religious entertainment (which limits its programming options), it also forbids advertising of alcohol, parties and other commodities and services deemed to be outside the Hindu religious scope.

Zelisa Boodoosingh-Rupani is part of the senior management team at the Jaagriti media organization. A former financial consultant, she is also a veteran of several radio operations in Trinidad and Tobago. Notbly, she has been a producer and on-air personality at local stations Heritage Radio and Radio Shakti, and has served as a media educator with local training institutes. From the graveyard shift on the Shakti startup, Boodoosingh-Rupani moved through the Indian radio industry and other local media in positions of increasing responsibility, and eventually joined the Jaagriti organization at its start. Among her roles in local media, she also recounted an attempt at religious outreach in the mainstream media, including television, where at one local station she pioneered a programme on Hinduism. At Jaagriti, in addition to her administrative role, she also hosts current-affairs programming on the radio and television operations. Boodoosingh-Rupani (interview by author, 16 June 2015) touted not only Jaagriti's success in overcoming local opposition to its existence, but also its success in transcending the limitations of its small environment:

> Regardless of all these struggles, we've stayed alive. We've been able to maintain ourselves. . . . I think what is phenomenal about Jaagriti is that we don't just touch lives in Trinidad and Tobago, we're not limited, but throughout the world – people from Iran, Iraq, India, Australia, Singapore – all over the world – Germany! We would get e-mails, and we've gotten e-mails from people who have really grown with us over the years. You know, they've stayed with us.

Additionally, Boodoosingh-Rupani (ibid.) explained this global reach in terms of what she perceived as the cravings that members of the diaspora may have for connections to traditions and culture:

> It would get the foreign and global attention, because these perhaps are the children who would have left Trinidad in the hope of something better, but they have kept something close to them, apart from their family, which would give them the connection by the radio station to connect, to extend greetings, perhaps, or to say

something nice, wishing them a good day . . . more importantly, keeping them intact in terms of religion, and what would have been close and dear to their heart that they may not have there.

Station administration did not appear to be actively pursuing the expansion of its foreign reach, particularly since it perceived the impact of the current streaming technologies as being more than adequate for its needs.

The persistent notion of Hindu identity being strongly associated with India has survived from the early political development of Trinidad and Tobago and remains a contentious part of the social and cultural landscape. On the question of whether a Hindu radio station represents something that is less than fully Trinidadian, as its detractors might suggest, Boodoosingh-Rupani (ibid.) argued that Jaagriti is Trinidadian because its content, while focused on an inherited cultural tradition, also attempts to preserve and protect the local expressions of that culture, arguing that as an organization it has "been able to inform, educate, but also keep the local culture alive in terms of the local, classical singing" and noted that "we do have a lot of local content with regards to what makes us Trinidadian". At the same time, this station administrator also pointed to strong connections with India, including visits from Indian personalities and officials, as well as connections for sharing programming materials.

The forced dichotomy between Indian and Trinidadian cultural elements does not necessarily pose a problem for the Jaagriti official, who also contextualized the particular focus on Hinduism as having its own Trinidadian character. Boodoosingh-Rupani (ibid.) noted that the particular evolution of Hindu philosophies in Trinidad and Tobago had its differences from Indian orthodoxies, and argued that the modern exchange of information in areas such as programming materials demonstrated both the similarities and differences in ideas and practices. Such dialectical tensions of tradition, place and time – articulated in history and negotiated in current (often globalized) social spaces – are all components of the complex identity claims and counter-claims that may seek to position the modern self. These many complex and interrelated factors are embodied in and dynamically engaged in particularistic and even (arguably) parochial media such as a Hindu-specific, Indo-Trinidadian radio station.

Conclusion

Radio Jaagriti's specialized focus on Trinidadian Hinduism demonstrates something of the diversity of the audience for Indian-format radio in Trinidad. Without the draw of either local chutney parties or Bollywood superstars, the station has managed to maintain its own audience segment and even expand into television production on local cable systems. Station management admits that with such limitations as restrictions on alcohol and cigarette advertising, revenues are perhaps not as high as they might be, but point out that it remains competitive.

Criticisms of this super-specialized station include the suggestion that by focusing on one religious tradition (and a particularly streamlined version of it), the station does not promote national engagement. Also, the particular version of Hinduism espoused through the station's content is strongly reminiscent of Brahmin caste prejudices and privileges that have permeated the local practice. The abrasive and sometimes reactionary political posturing of the station's parent body, particularly through its leader, has also contributed to reservations about the station in the broader mainstream.

CHAPTER 7

The "Corporate" Mainstream

The present chapter examines several Indian-format radio stations in Trinidad that share the common pedigree of being subsidiaries of existing business or mainstream media conglomerates. These four, belonging to major diverse business and media operations, include the second-oldest of the Indian format stations, as well the sector's newest entry. These stations are examined in terms of their content and their outreach, particularly as they wrestle with balancing the needs of their target demographics against their presence in an overarching national commercial media strategy.

While it is clear that even the founding of 103 FM was primarily a business venture, despite its many cultural trappings, there exists within the relatively small Trinidad and Tobago media market a group of Indian radio stations that fit more clearly into the mould of corporate enterprise than into the idea of independent radio. This was a distinction raised in conversations with radio practitioners in Trinidad and Tobago, though more often among those who sought to maintain an identity other than that of being a business solely in operation for the money.

Yet this was not necessarily a distinction that listeners to these stations seemed to be concerned about. In fact, listeners were more concerned about which stations played their favourite songs and featured their favourite on-air personalities. In several cases, the so-called corporate or conglomerate stations had several advantages, particularly being able to hire popular deejays and announcers.

The notion of the "conglomerate" station has to do with the existing structures of corporate media in Trinidad and Tobago. In a small media market there are limits to expansion and to the launch of media ventures. In particular, high-cost ventures such as television and radio have been risky and have been difficult to set up without existing support structures. When the government owned and controlled media, the resources needed would come from the national treasury. In a market-driven media environment there were no such subsidies, and start-up operations were limited in terms of available capital and other resources. One response to the constraints of market, capital and expertise was to build media operations that banked on existing corporate structures that might own and run newspapers alongside other corporate ventures, and then add radio and television operations.

Aakash Vani 106.5 FM

The name Aakash Vani (meaning "Voice from the Sky") hails back to the broadcasting services of All-India Radio, which was known as Aakash Vani since its inception in 1956. There is no little irony in the fact that this Trinidad station, using the moniker of the world-famous public broadcaster, is part of a corporate enterprise known as TBC (formerly the Trinidad Broadcasting Company), which is itself a subsidiary of a larger national corporate giant, Guardian Media Limited. Aakash Vani claims to be the number-one station for the subsector of easy-listening Indian radio, but is perhaps better known among its listeners for its popular morning programme, *The Morning Panchayat*. The term *Panchayat* (sometimes rendered as *Panchayte*) refers to a meeting, usually of elders or dignitaries, to discuss some matter of communal importance (K. Rampersad 2002, 103, 105). The station inherited this programme from its predecessor, Sangeet 106.1 FM. Contrary to prevailing stereotypes of Trinidad's Indian-format radio being isolated or insular, the *Morning Panchayat* programme has drawn both praise and criticism over its years of broadcast for engaging in national debates.

At its start, the programme featured personalities such as a former politician associated with the predominantly Indo-Trinidadian United National Congress, but eventually diversified in both its host composition and focus.

Aakash Vani started operations on 1 March 2007, with a focus on a slightly older demographic than some of its competitors. Among its audience draws are several of the well-established on-air personalities in Trinidad's Indian-format radio environment. Among the names, DJ Mamoo (Farouk Baksh) has been one of these leading personalities, having become established initially as one of the pioneering deejays of Indian-format radio at 103 FM. This shuffling of personalities within the small Indian-format media subsector has been a common occurrence as stations have moved in and out of the market and competition has increased.

This station also competes in the religious subsector of the market against Radio Jaagriti, because it provides nightly live transmissions of what are called *Ramayans*, ceremonial readings of sacred Hindu texts, often including music and songs. These presentations would usually be available to local communities, but have become national events (and even international events) with the provision of broadcast and streaming services from Aakash Vani.

A broadcaster at Aakash Vani, who asked not to be identified, described the station's audience as very dedicated listeners who sought solace and comfort in the station's musical offerings. By tuning in to Aakash Vani, they hoped to avoid what the broadcaster termed "the noise" of so many other genres such as rap and calypso and even some of the local Indo-Trinidadian content such as chutney. Even though most of the broadcasters speak no Hindi, the station also prides itself on collecting and sharing (via social media and on air) translations of Hindi lyrics.

The Superstation: Sangeet 106.1 FM

The older sister station of Aakash Vani is known as Radio Sangeet (meaning "music" or "a musical event"), or, as it presents its tagline – the Superstation Sangeet 106.1 FM. The station launched on 24 September 1995, some two years after the launch of 103 FM. Sangeet 106.1 FM was the first real competitor to 103 FM and enabled its conglomerate's programmers at TBC (what used to be the Trinidad Broadcasting Company) to break out of the limited-play Indian content that was the norm prior to 103 FM and to which Radio Trinidad 730 AM still held. It broadcasts from studios at St Vincent Street in the capital, Port of Spain, in a building that was formerly home to the *Guardian* newspaper.

A second full-time station catering to the Indo-Trinidadian market provided the grounds for competition on even terms. TBC was no longer limited to its traditional half-hour programming slots of Indian music, and, with Sangeet

106.1, began to push the boundaries of the format in order to compete with the by then well-established 103 FM. While some programming, including film music and religious celebration specials, mirrored the established 103 FM format, Sangeet 106.1 began to counter with the expansion of live coverage of community events. Its efforts in this direction eventually included such innovations as hosting "Chutney Cruises" by 2005 ("Chutney Cruise Ends with a Bang", *Trinidad and Tobago Guardian*, 10 August 2005, G33) and a "Chutney Parang Bandwagon" established by 2016, combining traditional Trinidadian-Spanish and chutney genres.

By the early 2000s, as the early Indian-format stations competed for market share, Sangeet 106 FM was also involved in experiments with trans-national broadcasts that originally involved telephone calls between deejays and stations in New York and Trinidad, but which eventually evolved into Internet streaming.

Sangeet 106.1 FM claims to be a leader in particular specific market segments and features a mix of chutney and contemporary Bollywood releases as well as remixes in which various combinations of digitized beat tracks, sampled music and sound effects are mixed to create reggae, soca or electronic dance music versions of Hindi or English songs. Its programming is aimed at a market segment of eighteen- to forty-year-olds. Whereas Aakash Vani broadcasts *Ramayans* at night, Sangeet focuses on chutney, boasting about its live broadcasts of major chutney events. Its involvement with the chutney and broader local music movements is evident in the presence of local chutney performers among the ranks of presenters on the air. Presenter Rick Ramoutar, for example, has been a frequent contender in the Chutney Soca Monarch competition ("Rikki Jai Ready to Defend His Crown", *Trinidad and Tobago Guardian*, 11 February 2012, 1).

90.5 The People's Station

Another local business conglomerate moved into the Indian-format radio market in 1997, soon after Sangeet emerged as a competitor to 103 FM. As part of what would become the media grouping known as the Caribbean Lifestyle Communications Network, 90.5 FM (with the tagline "The People's Station") launched on 17 August 1997. The station, which claims to have been the first to conduct live Internet streaming, also contends to have been the first Indian-format station to host a radio call-in show, entitled *The 90.5 Platform*, in 2001, though Sangeet 106.1 FM was broadcasting the *Morning Panchayat* from as far back as 1996.

During research in the early 2000s, 90.5 FM was found to be among the early stations streaming on the Internet and was observed handling requests from foreign callers and e-mails at the time (Mohammed and Thombre 2002). As with many of the other stations in Trinidad and Tobago's Indian-format radio subsector, 90.5 FM has been involved with several community events over the years. Among the most important of these is one that also demonstrates a wider national outlook, in contrast to prevailing notions of insularity or parochialism within the sector. Starting in 1999, 90.5 FM has hosted a national kite-flying event that has drawn on local traditions of kite-making. This event attracts hundreds of participants (more recently, including some from the surrounding island territories) and thousands of spectators "from every religious, ethnic and social group" (Seeta Persad, "Rain Holds Up Kite Flying at the Savannah", *Newsday*, 25 April 2011, 1). The eighteenth iteration of the festival was held in April 2017. Radio 90.5 FM also focuses on community and charity events, including Christmas promotions and giveaways.

The station has also distinguished itself with ventures into international linkages. Bearing in mind that the station has the support of powerful business interests and is therefore somewhat more insulated against financial risks, the management has conducted several ambitious projects with international dimensions. As has become the norm, the station has sponsored concerts by visiting Bollywood stars and playback singers, most notably Kumar Sanu in 2006 and Sonu Nigam, who performed in Trinidad in 2012 and 2014. However, beyond these relatively standard international ventures, 90.5 has also partnered with India's Zee-TV to broadcast the local Divali Nagar internationally, starting in 2015 (Paras Ramoutar, "Divali Nagar Live Broadcast: Radio 90.5 FM Partners with Zee TV", *Trinidad and Tobago Guardian*, 11 November 2015, 1), and has sponsored the Trinidad and Tobago cricket team in the regional professional league, aligning itself in the process with international Bollywood superstar Shah Rukh Khan (Anna Ramdass, "Radio 90.5 FM joins CPL Party", *Trinidad Express*, 27 June 2015, 1).

Taj 92.3 FM

The newest addition to the Indian-format subsector in Trinidad and Tobago is Taj 92.3 FM. This station forms part of a broader media operation with numerous radio stations and television operations known as One Caribbean Media. This group features several local radio stations playing adult contemporary as well as urban-format music and content, along with media operations in other regional countries, including Barbados and Grenada. The group's annual report for 2015 declared its total assets at over TT$888 million (approximately US$133 million).

Within these diversified media operations and substantial assets and holdings, the latest offering is Taj 92.3 FM, which started broadcasting in 2014. The numerous radio operations which include Taj 92.3 FM are managed on a broad scale by veteran radio personality turned media manager George Wayne LeBlanc. However, within the small and highly competitive Indian-format subsector, even management figures can become the subject of corporate poaching. This was evident in Taj 92.3's announcement in October 2015 that it had hired Indian-format stalwart Yaseen Rahaman (previously at 103 FM) to manage its operations ("Rahaman Is New Station Manager at Taj 92.3 FM", *Trinidad Express*, 7 October 2015).

In the crowded market for Indian-format music and entertainment, Taj has worked to implicate itself into both culturally specific events and those that are national in scope. In 2017, for example, Taj was the official station for the School's National Carnival Chutney Soca Monarch competition. The station also competes with more traditional events such as concerts with popular local performers and emerging talents from India.

Within Taj's programme offerings are traditional Bollywood standards as well as modern remixes and other examples of fusion. Its Friday-night and Saturday-afternoon offerings in particular tend to focus heavily on fusion mixes with host DJ Simply E (Gregory Prescott), who has worked previously with Masala 101.1 FM and Sangeet 106.1 FM, but specializes in reggae and electronic dance music mixes. Yet alongside such avant-garde content Taj 92.3 FM also features traditional stalwarts such as Rafi Mohammed, former producer of Indian-oriented television programming such as *Indian Variety* on TTT. This mixture suggests that the station programmes for the diverse tastes of the modern Indo-Trinidadian audience, which cannot easily be approached as a homogenous group.

Conclusion

The stations operating out of corporate and media conglomerates possess distinct advantages over those operating as independent ventures. With the ability to leverage market information gleaned from across their conglomerates' broad spectrum of media activities, these stations may be in a better position to programme for the diverse audiences within the sector and to meet the need to address the national audience as well.

Yet the complexities of Trinidad and Tobago society are such that even the corporate ventures have had to embark on diverse strategies to reach their intended audiences. In the case of Sangeet 106 FM and Askash Vani, operating from the same studios and under the same management, it has been a strategy of segmenting the local Indo-Trinidadian target audience by age and music preference, with one station focusing on chutney and party music for younger listeners while the other focuses on nostalgic Bollywood music and religious programming aimed at older listeners.

Other stations such as Taj FM approach the diverse audience segments with a wide variety of content including both nostalgic Bollywood classics and modern reggae-remixes on the same wavelengths. Yet as part of national media operations, these players also attempt to maintain a national outlook on their Indian-format stations. This is clear in the programming such as the *Morning Panchayat* programme, which deals with national issues and features a wide variety of guests and callers, as well as in events such as 90.5's national kite-flying competition.

CHAPTER 8

Hybrid Visions
WIN 101, Heritage and Shakti

While some stations have established a particular niche in the Indian-format subsector, this has not always proven to be the most commercially rewarding strategy. To reach the widest possible audience, many have had to broaden their focus. This has been more than a financial decision, however, as the multicultural and hybrid nature of Trinidad and Tobago often demands – as some of the key players in these stations have indicated – a hybrid vision of programming as well. In this chapter we examine several of the stations within the Indo-Trinidadian radio sector that self-identify as hybrid operations.

WIN 101 Masala Radio

Sunil Ramdeen was general manager at WIN Radio 101 FM until mid-2016, before the station went off the air over the non-payment of regulatory fees. Ramdeen is a former television journalist who established himself as a media personality in Trinidad and Tobago during his years with Caribbean Communications Network's TV6, which launched in 1991. As general manager of WIN Radio 101 FM and its companion television station WIN TV, Ramdeen maintained an important on-air presence, particularly in news and current-affairs programming, where he hosted important national debates on political and social issues.

Ramdeen (interview by author, 16 July 2015) credited the efforts of the founders of 103 FM with opening the doors to the establishment of various Indian media companies in Trinidad and Tobago, noting that the existing mass media houses at the time resisted efforts to start the first Indian radio station. He spoke of his encounters with WIN Radio 101 FM founder Mohan Jaikaran and how the power of the local market drove the establishment of Jaikaran's station, building on 103 FM's market success: "103 went on to be hugely successful, and they're still very successful today, and it opened the door for everything else. The fact is, there's a huge Indian community that simply wasn't being catered for unless it was Indian Arrival Day or it was Divali or Eid."

Ramdeen (ibid.) said that his experience with the (now defunct) HCU's efforts at setting up radio and television stations demonstrated the strength of the Indo-Trinidadian market, and his contact with music producer and promoter Mohan Jaikaran led him to WIN 101 FM and WINTV. Ramdeen's description of the initial focus of the station is instructive as to the dynamics and challenges of media planning in a small, fragmented market:

> Our original business plan was for Indian or high Indian content. Mohan brought in some business partners from the United States who changed that, and they went with time-worn, they went with traditional programming. We found that we were one of four national stations fighting for the same market share. Advertisers would come to us and say, "Who are you catering to?" And we couldn't really give them a direct answer, and sometimes when we did, it was sort of like: "Oh, well, we have that already." The minute we switched to Indian TV we didn't have to answer that question from advertising any more; they knew very well who we were. Indian entertainment, chutney and so on, exploded at that time – it became really big – and not just locally, internationally as well, because of the millions of diaspora members.

WIN Radio 101 FM emerged in part out of an earlier effort known as Masala 101.1 that started operations on 31 August 2000 as a subsidiary of the HCU. The term *masala* refers to a mixture (usually of spices). Using this notion of a mixture (and using business reasoning), Ramdeen (ibid.) argued that the all-Indian format could coexist with more diverse offerings, questioning, in this argument, something of the misplaced essentialism involved in the assumption that an Indo-Trinidadian is limited in his or her scope of reference only to Indian content:

> The thing is, if you have an all-Indian network, you limit your demographic. For instance, I'm a young Indian person, that doesn't mean I'm only going to listen to

> Indian music. So if you're catering to everybody, I will sit down and I will listen to reggae, I will listen to soca, I will listen to dance [music]. So if we are catering music for, and to connect with, young Indian people, we can't just play Indian. They're not going to listen to us.

Here Ramdeen made some important distinctions among the audience groups and the extent to which age was a factor in the propensity to respond to mixed content:

> You know, forty-five, from a media point of view, those are the people who spend money. Those are the people that the advertisers want to get. If we want to get the young people, fifteen to forty, to listen to us, you need to go after that. Factors in surveys that have been done have proven that 101, in the age group fifteen to thirty-five or forty, is the number one Indian news station. Overall, 103 may get a rating of one, but a great of part of their listenership are over forty-five, over fifty.

Ramdeen (ibid.) noted, as well, that the audience has increasingly begun to reflect global and regional forces that extend its frame of reference beyond the traditional Indian nexus, resulting in changes in programming across the sector:

> If you listen to local Indian radio . . . they are all changing their format to reflect 101's format. 106.1 FM has tried to completely copy our format. 103 has started playing music they've never played before . . . even some soca, somebody told me. So they are recognizing that you need to cater to all types of music, because you can't say because you are [Indo-Trinidadian] you only listen to Indian music. We all listen to a diversity of music. All we do is give Indian music the place it deserves. It can't be relegated to one song an hour.

However, even with the Indian component of that diverse mix of music, Ramdeen (ibid.) cautioned that about 80 per cent of the "Indian" music on his station was locally produced material and about 20 per cent imported from Bollywood. Additionally, Ramdeen noted that the global scope of WIN 101 was always important, with its founder considering it a "world music" platform, which remained the key motivation for the station's expansion into the North American market:

> I remember when it launched, it launched as world music, and that was the first, because Mohan's view was always a world view. He always created content for here and the external market. That was always the dream. That is why we got into Canada;

that is why we are getting into North America. Before he died he had talked to Time Warner and Comcast, which [talks] are continuing. So it was always global . . . So we were then able to really refine our content to reach both the global market and to appeal to the effects in our work.

Heritage Radio

When Hans Hanoomansingh launched a radio station in 2005, given his history of broadcasting Indian content in mainstream media, his involvement with 103 FM and his role in the National Council of Indian Culture, a common expectation was that his station would be another Indian radio station. Heritage Radio turned out to be very different from the stations that had, by that time, defined Indian radio in terms of content and audience appeals – defining itself, instead, in terms of the broader set of influences that exist in the sociocultural mix of Trinidad and Tobago. Heritage FM broadcasts on 101.7 megahertz from Woodford Street in Port of Spain (though it started operations in the south of Trinidad, in San Fernando). Programming includes not only Indian music (including a new incarnation of Hanoomansingh's successful *Melodies of India*) and live *Ramayan* broadcasts, but also live services from Christian churches and a wide range of local music and programme types.

Hanoomansingh (interview by author, 15 July 2015) described his thinking around this station in both cultural and broadcast business terms as follows:

I rationalized that there were enough Indian stations in the country. So I will do something that's different. I would celebrate all the culture of Trinidad and Tobago.

I had already done some things in the Indian community, like, for example, I started *Ramayan* (Hindu religious scripture readings) broadcasting. I started *Khutba* (sermons) from the mosques every Friday. And with a lot of support from the various organizations I started twenty-four hours of Divali programming, thirty days of Ramadan programming and so on – so, a fairly comprehensive platform for the expression of Hinduism and Islam in the society.

So can I retain that, but also celebrate on Heritage Radio [Spiritual Shouter] Baptist Day, and Easter and Good Friday and Christmas and Independence and Republic Day . . . So that was the vision, and . . .for example, when we started Heritage Radio we also included a calypso programme called *Real Kaiso* . . . We have a Latin programme called *Conexión Latina*. So yeah, we've gone a slightly different way in our programming concept.

Hanoomansingh's inclusive approach to programming reflects not only the imperatives of multiple cultural forces in Trinidad and Tobago but also an awareness of the broader global forces on the medium and its content. Hanoomansingh (ibid.) connected this global awareness with continuing (and often increasing) contacts with India, commenting:

> There's a relationship with India that I've had. For example, the first broadcast, live actuality report I did, that won the prize for Radio Guardian, was at the memorial for the second prime minister of India, who died while on an official visit to Tashkent. And apart from music I introduced documentaries related to the Indian struggle for independence. I have maintained that kind of interest. And then I led the National Council of Indian Culture for twenty-five years, so there's been a relationship.

In further testament to the relationship with India, Hanoomansingh has also expounded on the Indo-Caribbean experience as an invited speaker at the Indian Government's *Pravasi Bharatiya Divas*, the largest assembly of people of Indian origin in the world.

Yet, for all these connections, Hanoomansingh does not limit the scope of his programming either culturally or in terms of form. He dismissed the view of modern radio as a kind of jukebox playing a rotation of favourite songs for listeners, and particularly one that plays only one kind of song. During our conversation, for example, Hanoomansingh indicated that there were discussions about including a programme of African music on Heritage. He also pointed out that the challenge of making programming that is relevant to a national audience requires constant work and adjustment:

> The reality is that we are still contained with narrow domestic walls. So there are people who would hear Indian music on Heritage Radio and say, "That's an Indian station, I'm not going to listen." And the converse is also true – they'll listen to WACK Radio in San Fernando and hear steelband and calypso. It is the same. We've tried and we cater for a listener who has an appreciation, who understands that even . . . 103, for example, [has] 100 per cent Indian music, but the introduction of the songs is in English. So we have our limitations in terms of the language.
>
> Ravi Ji . . . who in his earlier years was very involved in calypso and steelband and so on, he says: "You know, I listen to 103 all the time. I have one regret. Because I listen to 103 all the time I don't know the road march at Carnival time. Because 103 will not be catering for that." So we try to be reflective of the society.

Hanoomansingh also pointed to the complexity of the cultural interrelationships within Trinidad society over time, noting, for example, that the same people who would call on him to present shows of nostalgic Indian songs also grew up listening to him play the Beatles and Elvis on the radio.

Radio Shakti/Hot Like Pepper Radio

Radio Shakti (often translated as "power", or "empowerment") was one of the media operations subsumed under the now-defunct HCU during the early 2000s. A complicating factor in this station's emergence and persistence was the fact of its relatively unusual situation while operating under the HCU. Parliamentary reports stemming from the HCU failure indicated that the parent company of the Shakti operation, Upward Trend Entertainment Ltd (which obtained a radio broadcasting licence from the state in 2004) had entered into agreements with the HCU such that a subsidiary of the credit union, called HCU Communications (headed by veteran media personality and Indian-format stalwart Hansley Ajodha), would operate on a licence owned by Upward Trend, with the option for HCU to purchase 75 per cent of Upward Trend for $TT5 million (approximately US$750,000), which did not materialize despite a 10 per cent deposit on the deal being paid in 2000 (Colman 2014).

Under its arrangement with the HCU, Shakti found itself embroiled in several political and legal firestorms. An interview with former prime minister Basdeo Panday on Shakti, for example, was thought to be one of the precipitating factors in attacks on the credit union and its leader (Yvonne Baboolal, "Harry: Don't Blame Me for Depositors' Plight", *Trinidad and Tobago Guardian*, 11 July 2012, 5) while, on more than one occasion, comments made on Shakti regarding the handling of the HCU receivership led to lawsuits claiming that the HCU leadership had engaged in defamation (*Robin Montano vs Harry Harinarine*, 2012; Sandra Chouthi, "Big Worry for HCU Depositors", *Indo-Caribbean World*, 20 August 2008). Even the arrangement under which the HCU was engaging in a broadcasting operation using someone else's broadcast licence came in for questioning from local detractors (Yvonne Baboolal, "Radio Shakti under Upward Trend's Wing", *Trinidad and Tobago Guardian*, 17 August 2008, 5). Here, we are less concerned with Shakti's troubled history and more concerned with the evolution of the station as an independent operation after the collapse of the HCU, starting in 2008.

In the years following the collapse, Radio Shakti moved from the HCU's

Chaguanas offices to facilities in Couva. There, the broadcasting identity changed from Radio Shakti to "Hot Like Pepper Radio". This title reflected the history of the Upward Trend owners, Anand Rampersadsingh and his wife Ingrid Rampersadsingh, who in 1998 launched a programme entitled "Hot Like Pepper Indian Radio" on an ethnic radio station in Toronto, where they lived at the time (Kayum 2008). This was followed by the establishment of an AM frequency known as "Hot Like Pepper Radio" on 530 AM, operating briefly from 1999.

In its Trinidad and Tobago incarnation, the "Hot Like Pepper" station, without the financial backing of the HCU, emerged into a much more eclectic station, not only in its programming, but also in its outreach and business efforts. In this regard, Hot Like Pepper Radio has called itself the "only multicultural radio station in Trinidad and Tobago" ("Divali 2016 Packages", Facebook post, 1 October 2016), claiming to capture a significant audience of different ethnicities and religious backgrounds throughout Trinidad and Tobago. While the station involves itself with certain mainstays of the Indian-format sector such as live party events, it also focuses to some extent on traditional forms, such as what is known as "classical" singing. The station has, for example, positioned itself as a promoter for a ladies' classical singing competition and features a Sunday-morning segment for local classical singing. Yet this station also features promotions and production efforts for numerous other cultural events outside of the Indo-Trinidadian orthodox tradition. Saturday and Sunday mornings even feature several slots dedicated to gospel music and content. Its promotional materials include posters with the tagline "Moving Forward as One" that feature photographic mosaics including Bollywood singers Asha Bhosle and Lata Mangeshkar alongside calypsonian the Mighty Sparrow.

Conclusion: The Hybrid Context and Entrenched Globalisms

The hybrids noted here are later arrivals compared to the initial efforts of 103 FM, which defined Indian radio in Trinidad. To clarify, the WIN 101 operation evolved out of an earlier station that attempted a hybrid approach as well (described as *masala,* or mixed) but was eventually sold. WIN 101's fate at the time of writing is uncertain and this may be in part due to economic factors or a combination of such factors with poor decisions on the part of the ownership with regard to its regulatory obligations.

While the notion of hybrid stations seems a logical choice in a society replete with cultural influences and global pressures, it remains unclear whether the hybrid approach fits with the media habits of audiences in Trinidad. As Hanoomansingh pointed out, one listener may switch away upon hearing Indian music, while another may switch away on hearing soca. However, as Ramdeen noted, no listener in Trinidad listens to only one kind of programming, and younger Indo-Trinidadian audiences are avid consumers of soca, reggae and American pop music.

In some senses all the Indian radio stations in Trinidad are, in fact, hybrids. They feature songs from Indian movies played for audiences who, by and large, do not speak the Hindi language of those songs. They also play "local Indian" songs which may be in Trinidadian-dialect English or Trinidadian "Hindi". Also included among their selections are mixes in which Hindi songs are set to Caribbean rhythms. In all these ways, even those stations claiming to be Indian stations exemplify hybridity in its many forms.

One employee at a popular "Indian" station who spoke on condition of anonymity expressed a slightly more purist view of the tendency towards hybrid content, describing the inclusion of soca and calypso with Indian music on some of the stations as "jarring" and "disquieting". However, even absent the soca mixes or the inclusion of American pop, soca or reggae, these Indian stations cannot avoid hybridity, as they are all hosted in English. As the same interviewee indicated, translations of the meanings of Hindi songs for the non-Hindi-speaking Trinidad audience are an important part of the role of the announcers on all Indian stations.

Clearly, the complex intermingling of influences in Trinidad society and in the programming of the overtly hybrid stations (as well as that of their avowedly "Indian" counterparts) demands the application of a broad concept of hybridity, as noted previously in the work of Kraidy (2005) and Pieterse (2001). The multiple competing influences of Trinidad and Tobago's particular cultural mix include diverse commingled elements that cannot be easily separated into component threads.

The particular juxtaposition of the overtly hybrid (for example, stations that call themselves "mix" or "*masala*") and those that lay claim (however implicitly) to some kind of purity demonstrates the existence of a deeper debate within the culture. It is one that exists in all cultures with multiple influences – the debate over cultural maintenance and cultural assimilation. The purist position voiced by at least one interviewee that considers soca music to be intrusive in Indian programming, for example, attempts to isolate and protect the one cultural element from the other. The assimilative position, voiced by Ramdeen and to some extent Hanoomansingh, acknowledges the vast and inevitable interplays of hybrid forces to generate new and interesting hybridities.

Yet here as well, one must ask the question that Pieterse (2001) poses – so what? Does the notion of hybridity really produce any insight into cultural processes except to make the rather obvious argument that cultures in contact will intermingle? What is also obvious in the evolution of Trinidad's Indian radio, in both its mainstream and niche forms, is that the notion of a hybrid as a fixed entity is not applicable. The constant evolution of forms such as chutney and their interplay with Indian radio and the sudden introduction of Indian/ Caribbean mixes are examples of the lack of fixed form that characterizes many cultural phenomena and argue against the fixedness with which one may conceive a proper hybrid.

Thus hybridity, here in its broadest form, is a convenient label for some of the developments in Trinidad's Indian radio, but does little as an explanatory framework. It may be more useful to refer instead to the deeply entrenched historical and global influences that have shaped Trinidad society for a better understanding of the "hybrids" that one may find there.

The global historical forces that delineate modern processes and tensions of cultural expression and cultural politics are, of course, not limited to the Indo-Trinidadian radio context. Within the broader society, as in the mass media of Trinidad and Tobago, several entrenched globalized influences have jostled for prominence over the years. During the 1970s, post-independence identity discourses demanded interrogation of the assumptions surrounding media content. With television production in its infancy, and with a thriving local music environment, local radio became the focus of this discourse, with several camps, including music producers and performers, arguing for a greater focus on local music and airtime quotas to guarantee parity with foreign songs (Meschin 2004). While this discourse did not generally include acceptance of Indian music (even local compositions) as local, it did raise questions about the interplay of global influences on the radio. This debate continued into the twenty-first century. In an interview with *Billboard* magazine in 2004, for example, musicians and producers from Trinidad and Tobago claimed that local radio stations dedicated only about 5 per cent of their airtime to local music (ibid.).

This debate over local versus foreign music fits into a neat and acceptable convention of a binary relationship between dominant foreign media and their usurpation of the legitimate place of "local" or indigenous fare. Such a simplistic view (common in the era of Schiller's "Media Imperialism" thesis and the drive for a new world and communication order) overlooks the complexity of the identity claims implicitly being made. The claim of "local" is itself flawed in this debate, excluding, as it traditionally has, local material from ethnic groups outside the Afro-Trinidadian mainstream. More importantly, however, the notion of "local" conveniently ignores the numerous global influences present in Trinidad and Tobago's local music and its Carnival celebrations, where much of that music originates. These influences include both African and European currents (Liverpool 1998, 2001) and countless evolutions and mixings of these and other flows (Sofo 2014), along with the rhythms of Latin American music.

Thus "local" music, in the forms of soca and calypso, is itself a particular hybridized form drawing on global currents. Such an argument is made even clearer with more recent concerns about undue influence of Jamaican dancehall reggae on Trinidad and Tobago's "local" music.

Even the so-called traditions of Trinidad's Carnival and "mas" (masquerade) are themselves cultural aggregations of a variety of influences. Franco (2007, 28), for example, notes how that which is assumed to be "traditional" may really be pragmatic:

> In 1950s Trinidad, the nationalist movement required cultural icons that expressed not only the changes and new face of the nation but also represented the cohesiveness of the national community. Carnival, as the most prominent of all cultural forms, quickly became the icon par excellence because, presumably, it was a very public display of nationalist ideology. It was in this nationalist climate that the traditional characters were imagined and created.

Nurse (1999) also extended the debate on globalizations by challenging prevailing notions of wealthy dominant nations exercising one-way cultural influences on smaller emerging nations. In Nurse's analysis, various elements of Trinidad and Tobago's Carnival act upon cultural spheres in North America and the United Kingdom in such a way as to exercise a kind of "globalization in reverse" (ibd., 663). This counter-flow can more broadly be taken to include

Caribbean influences on US popular music dating back to fears about calypso eclipsing rock and roll in the 1950s (Funk and Hill 2007).

Thus outside the sometimes awkward identity claims of Indian radio (with its concessions to hybridity on the one hand, and purists claiming some kind of superiority for Indian cultural forms on the other) the mainstream Afro-creole cultural forms are also subject to interrogation. When the assumptions of indigenousness and the illusions of historical fixedness are removed, the foundational aspects of the mainstream "local" culture are rendered much less stable. In that context, the negotiability, fluidity and hybridity of the entire society in historical and global contexts are much more evident. The stamps of historical and embedded globalizations become evident and the fallacies of dominant cultural narratives are exposed as chimeras.

Indian-format radio in Trinidad and Tobago brought issues of identity, authenticity and hybridity into focus by foregrounding not only Indo-Trinidadian cultural identity but also by bringing into question deeply entrenched assumptions about the cultural mainstream and the negotiation of cultural spaces. The Afro-creole mainstream has also been affected by the Indian-radio format, particularly as traditional boundaries, such as that between chutney and soca, are being redefined, with Indian-format radio as a nexus for these negotiations. As we have seen above, as well, the Afro-creole mainstream has also been forced to encounter Indian-format radio as a business necessity, with mainstream media operations venturing into Indian radio, even if only for the economic benefits of doing so.

CHAPTER 9

The Global Dimensions

The global dimensions of Trinidad's Indian-format radio sector operate simultaneously with its domestic broadcasts. Live Internet streams and social media provide some means of unification for families, including so-called transnational families with members in Trinidad and Tobago as well as members who have since migrated elsewhere to places such as North America. Family members may be united through listening to the same content at the same time and also through the seamless delivery of greetings and song requests among them, irrespective of their physical location.

Indian-format radio in Trinidad and Tobago began as a parochial undertaking within the context of a small national broadcast market of just 1.4 million people. Technical developments in media and connectivity soon made these efforts accessible to audiences outside the national market, and a modern history of migration from Trinidad and Tobago had already created diaspora communities in North America and elsewhere. Well after their permanent settling in Trinidad and Tobago, Indo-Trinidadian families also began to experience out-migration as educated family members sought opportunities in the United Kingdom, Canada and the United States. Although some authors such as Teelucksingh (2011) have focused on the so-called Black Power movement of the 1970s and crime in the 1990s as push-factors for Indo-Trinidadian migration, there are clearly many other factors involved. Among those other factors were the changing relationships among Britain and its colonies in the 1950s and 1960s, and the increased range of opportunities for economic and social advancement offered in larger, more developed countries – which continues to be the primary driving factor for Caribbean out-migration (S.N. Mohammed 1998). For whatever reason, Indo-Trinidadians today often have relatives in the United States, Canada (and the United Kingdom to a lesser extent). These "secondary" or "new" diasporas form part of the potential markets for Trinidad Indian radio – although it was not clear at first how these markets would materialize.

Ethnic media use among Indo-Trinidadians prior to the launch of the first Indian radio station did in fact recognize an external scope. Long before the existence of Indian radio stations, the Internet or audio streaming, local listeners to short Indian-music segments would write letters to the hosts of these segments sending greetings or requesting songs to be played for their friends and relatives, including their relatives abroad (Mohammed and Svenkerud 1998; Mohammed and Thombre 2014). An observer would, in those days, perhaps question how this system was meant to work, as there was little chance of a relative in the United States, Canada or the United Kingdom receiving the radio-station signals, which were weak enough to be sometimes difficult to receive in Trinidad. Audience members placed significant value on these greetings and requests and on their families being mentioned on air. Thus they often developed mechanisms by which these long-distance dedications would come to the attention or even the ears of their relatives. One such mechanism was simply to inform the relative by letter that a request had been sent and a song played for them in Trinidad. Later, when recording technology became available (and this strategy persisted into the early days of Indian radio stations), the segment in which the request or greeting was announced would be audiotaped off the radio broadcast and either saved for a visit from the family member or even played over the phone during a phone call (as several listeners described).

This awareness of the external audience was not far from the minds of the earliest operators of Indian radio stations. Dik Henderson (interview by author, 15 July 2015) noted that soon after the launch of 103 FM, the station garnered international attention among like-minded broadcasters abroad:

> There were people all over the diaspora, in Canada and North America and England, who were buzzing about it. . . . In the early days we had all different kinds of approaches from people to do link-ups with us . . . but the thing is, we didn't have the technology at the time. Now everybody has the technology, people are doing their own thing. But yeah, people were very, very keen to join up with us. In Suriname . . .

there was a lady with four radio stations who wanted to team up with us . . . but we couldn't speak the language. Meanwhile we were with Guyana and so on . . .

Soon after the launch of 103 FM and well into the early years of Trinidad Indian radio, its global audience market potential also began to show as diaspora-based Indo-Trinidadians became part of call-in requests and stations began to feature specials broadcast from New York and Canada. At first, these transnational engagements were sporadic and complicated, particularly before the Internet facilitated audio streaming, using some of the methods described above. However, early in the popularization of the Indian radio stations, their management and technical staff attempted to create broadcast events that were global in scope using existing technologies. Sometimes the broadcast would feature callers from abroad, or a guest deejay hosting some segments via a long-distance call. There were several instances of dual simultaneous broadcasts with stations based in the United States and those based in Trinidad connected by phone lines and taking calls from listeners at both ends. These were sporadic and expensive, and were to soon be replaced by much more cost-effective and far-reaching audio-streaming technologies.

Streaming

The small size and fragmented market for Trinidad and Tobago media, as well as the market potential of the diaspora, prompted many Trinidad Indian-format radio stations quickly to adopt Internet-based audio-streaming technologies in the early 2000s. As with many developments in the history of media, there are multiple (often competing) developments that led to audio streaming, and it is often difficult to establish definitive timelines and firsts. Modern media forms also feature numerous layers of technology that fit seamlessly and often invisibly into the process of delivering content.

Radio itself emerged out of developments in the understanding of the electromagnetic spectrum, electricity and wave propagation. In its earliest incarnations, the ability to transmit the atmospheric interference from a spark gap from one location to another (at first, not very distant) location was not a revolutionary premise – at least not as far as the Italian authorities were concerned. In fact, Marconi was unable to elicit a response from his own government and at least one source has suggested that officials considered his proposals worthy of an insane asylum (Solari 1948). Thus Marconi ended up going to England in 1896 (Hong 1994) to pitch his "invention", which was still not radio when he promoted it to British officials. While traditional wisdom suggests that his early efforts involved an approach to wireless telegraphy, there is evidence that even this might not be true, as Garratt (1994, 75) has suggested: "Marconi's original approach was made not to the Post Office but to the Secretary of State for War. The invention he was offering in his original letter . . . has turned out not to be for a system of communication but for a system for the radio-control of torpedoes or other unmanned vessels!" What we understand as radio would not evolve until several iterations of the technology were tested and used, primarily for maritime communication, where traditional wire-connected telegraph technology was not feasible. This was the wireless telegraphy that is widely known as the technology's first use, but it was limited to transmission and reception of Morse code dots and dashes. Evolution of the technology beyond its primitive binary form would require the work and imagination of the Canadian Reginald Fessenden over a period of several years, culminating (to an extent) in 1900, as Brodsky (2008, 1866–67) has recounted:

> When Fessenden learned of Marconi's spark technology he immediately recognized its limitations. Sparks are inefficient, invite interference, and are ill-suited to carry speech. Fessenden knew exactly what the problem was, and it wasn't long before he conceived a solution. Sparks produce "damped waves" – electromagnetic waves that fall rapidly in strength. . . . Fessenden envisioned a simpler and cleaner technology. He knew that if he could produce steady electromagnetic waves. . . . Fessenden demonstrated his genius when he modified a spark transmitter to transmit marginally intelligible speech. It wasn't by any stretch of imagination a marketable solution, but it proved that speech could be sent over wireless. . . . Reginald Fessenden transmitted speech wirelessly for the first time in history on December 23, 1900. The transmission took place on Cobb Island over a distance of one mile. His voice was badly distorted and accompanied by a loud noise. Reportedly, Fessenden's first words were "Hello, one, two, three, four. Is it snowing where you are, Mr Thiessen? If it is, telegraph back and let me know."

Radio then took over two decades more to become a medium for the broadcasting of voice and music and several more years to develop into a commercial and international medium for information broadcast widely to listeners. Audio-recording technology developed alongside broadcasting and several methods of capturing and reproducing sound became available to audiences during the early twentieth century. Phonograph records made of various materials, including wax composites and vinyl, endured as a medium of primary importance for consumer sound even as tapes competed as a more versatile alternative. While these media dominated the consumer environment, scientists and engineers were already working on the means and methods of digitizing audio signals as well as the ability to compress the data thus generated into manageable-sized files. The consumer audience would not have broad access to these technologies until the advent of the compact disc in the early 1980s. These discs provided improved sound quality compared to the existing alternatives tapes and records, driving their diffusion and adoption (Rogers 1962), but the underlying power of digitized music would not become evident until a confluence of technologies made music on the computer both viable and attractive to consumers.

Partly owing to the popularity of the free file-sharing system known as Napster starting in 1999, but also predicated on the diffusion of the personal computer starting in the early 1980s and the spread of the Internet among consumers from 1991, the sharing of so-called MP3s (digital audio files incorporating the Frauhofer compression algorithm denoted as MPEG-3) marked the start of a revolution in digital audio. Once audio could be created, manipulated and shared on the computer, users began to engage in a variety of production and consumption activities. At first, limited bandwidth and slow speeds meant that content would be shared by download and the user would have to wait for a substantial time for a sizeable audio file to completely transfer to the local computer before listening to it. Despite these limitations, in 1993, Carl Malamud launched a downloadable audio production under a project ambitiously called "Internet Talk Radio". O'Brien (1993, 96) described the effort as follows:

> Carl Malamud, author of "Exploring the Internet: A Technical Travelogue", has developed Internet Talk Radio in conjunction with O'Reilly & Associates publishers and Sun Microsystems. The program consists of half-hour audio files that will be spooled on the Internet for users to play on individual PCs, as well as on Macintoshes and RISC-based [reduced instruction set computer] workstations. . . . The "radio" begins its programming at the end of March with a series of interviews called "Geek of the Week".

The notion of streaming would require several more years of innovation, with an untidy slew of technologies aimed at compressing audio and delivering it rapidly over limited bandwidth. Systems such as those of RealPlayer and various Windows Media streaming offerings aimed at rapid compression and delivery of data prior to full download, with a small buffering period of initial acquisition. Added to the mix of compressor/decompressor (CODEC) battles were choices about how streaming should best be achieved. Setting up and maintaining streaming audio (and later video) servers seemed simple enough in principle at the start, though operators rapidly realized that questions of scalability and costs began to arise. For small operations, the additional hardware, software and human-resource burdens of implementing streaming soon seemed to suggest farming this function out to specialists rather than investing in in-house capacity.

As early as 2000, radio broadcasters in the United States had already begun to pursue Internet streaming as an essential component of their operations. At that time, *Music Business International* ("Radio Broadcasters Look at Taking Business On-line", 1 April 2000, 31) noted that in the United States, "for traditional radio broadcasters it is no longer a question of if, but how they are going to incorporate the Internet into their strategies", noting, however, that expenses and uncertainty were barriers at the time and that while many US radio stations had "grasped the webcasting nettle with enthusiasm and imagination", just as many were still "hanging back, reluctant to take what they regard as an expensive leap in the dark". Despite these reservations, most radio broadcasters in the United States, faced with the emerging technologies, including the increasing availability of both the Internet and broadband connectivity, recognized the necessity of incorporating streaming into their operations. Industry monitors were tracking the numbers of stations streaming worldwide since at least 1997, and January 2000 numbers revealed that there were in excess of three thousand radio stations streaming on the Internet, with 48 per cent based in the United States and Canada and 42 per cent being international stations moving their local programming online (ibid.). These developments seemed to provide a direct and smooth path to the business of streaming. However, there was no guarantee that the business models of traditional media would be the way of the future, so that new forms of media were beginning to spring up around the new technologies as well.

A service aimed at Indian immigrants in the United States operating out of San Jose, California, started up with the promise of bringing India to those in the diaspora. The company, calling itself Homeland Networks, offered early versions of what is commonplace today, media streams (described then as bringing "real-time multi-lingual, multi-channel streaming radio and TV to the Internet") that included brands such as RadioofIndia.com (ROI), TVofIndia .com (TVOI) and PressofIndia.com (POI). Regarding these developments, *India Currents* ("Bringing India to You", 31 March 2000, 24) commented:

> It is a great time to be an Indian immigrant in the U.S. Although half a world away, never before has he been closer to his roots. Telecommunications and the Internet have amazingly shrunk his world. . . . Even though we learn to live around it, there are sounds and sights from the native land that we crave, like that soulful Tamil melody or *bhajan* heard on the radio. Or watching an old Hindi movie on TV every Sunday afternoon. Or sitting on Grandpa's lap and reading the Gujrat Samachar. . . . ROI provides 24/7 streaming of news, sports, and regional music from India while TVOI features sports and entertainment. A highlight of both is the live broadcast of cricket matches. With 164 regional, national, and world newspaper and magazines, POI links and drives traffic to specific sites and acts as a newsstand that can be customized.

The eventual failure of this and other grand ventures of the dotcom boom, including high-profile operations like Broadcast.com (which Yahoo purchased for $6 billion), created even more uncertainty for the media environment, which proceeded somewhat more slowly and carefully into the realm of streaming media in the years to follow.

Meanwhile, the highest growth at the time was in small local stations around the world beginning to capitalize on the reach and potential of the Internet for delivering media by migrating online. Trinidad and Tobago was no exception, and by 2002 three of the four Indian radio stations in Trinidad (103 FM WABC, Masala 101.1 and 90.5 FM, operating as Radio Central) engaged in some kind of regular streaming of content (Mohammed and Thombre 2002).

Global Technology and Global Culture

Both the India-Trinidad and Trinidad-North America cultural links associated with Trinidad Indian-format radio prompt attention to the impact of new global media technologies (here in combination with traditional radio), including their roles in creating and maintaining cultural identities among geographically separated communities (Adams-Parham 2004; Georgiou 2006; Hiller and Franz 2004; Horst 2010; Mohammed and Thombre 2011). Particularly relevant to the local and global operations of Trinidad Indian-format radio stations is Miller and Slater's (2000, 6) argument that "modern nations might be thought of as 'imagined' or 'virtual communities'", dependent on the capacity of media to "reflect a singular imaginary back to a dispersed or divided people". They argued that "this is particularly apt in the case of Trinidad, which has had to imagine national and cultural identity across a complex ethnic mix and a geographical dispersion across the globe".

This complex mix includes an increasing amount of exchange between the Indo-Trinidadian community and India. Indian radio is no longer the only source of Indian-based cultural material. Such links now include trade expositions and several commercial establishments for the sale of Indian goods. Indian movies are also no longer under the control of a select group of distributors, but available online as well as on several cable channels. The impact of these linkages and exchanges on Indo-Trinidadians (most of whom do not speak Hindi or any other Indian language) who have been integrated into the broader culture of Trinidad and Tobago over several generations remains to be studied. Existing assumptions about the levels of cultural retention (Samaroo 1987) and the role of mass media (Malik 1971), as well as the notion of cultural loss through assimilation or "creolization" (P. Mohammed 1988; Sampath 1993) may be overturned in favour of a more post-culturalist approach in which users may choose and adopt practices from available options in unique combinations.

The Trinidad Indian-format radio case is not unique. Many other examples exist in which global media technologies are used to connect historical primary-diaspora communities with modern secondary-diaspora communities that include Jamaican, Haitian, and other groups (Adams-Parham 2004; Georgiou 2006; Horst 2010). In all these cases, particular global influences have forged identifiable cultures that have given rise to diaspora communities that display properties of both their home countries and their ancestral roots.

Officials at 103 FM indicated that they plan to capitalize on the global reach of Internet technologies and the diverse opportunities that digital radio now provides, though they note that their core audience remains the listeners in Trinidad and Tobago (Henderson, interview by author, 15 July 2015; Rahaman, interview by author, 15 July 2015). On the question of the internationalization of their markets, 103 FM managers described their global operations as only partly the result of business and technology imperatives, adding that audience participation in several forms also acted as a driving force. According to Saith (interview by author, 15 July 2015):

From being one of the announcers on the station, I could tell you that part of the reason that we started [Internet] broadcasting was because we had calls, we forever have been taking calls from listeners. And we had, especially on the weekend . . . people calling: "Hey, I'm calling from New York, or Canada . . . Hail out my family here!" And even before we started [Internet] broadcasting . . . there was this gentleman at [the website] Trini Radio, I think, who used to record our programming and play it back later. And that was at the very, very first stages of broadcasting. He used to record on a DAT [digital audio tape] for hours, for four or five hours, and broadcast at some other time to satisfy that audience up there. It was very small then, but that's how we started.

The management of 103 FM estimated its total hits on its streaming service at about three hundred thousand per month, with the Trinidad and Tobago audience accounting for the majority of streams, but the United States and Canada being the second most popular sources of listeners. They also indicated some listenership from the United Kingdom, the broader Caribbean, India and China. As with other Trinidad Indian radio stations, 103 FM also encourages vicarious participation by streaming the studio feed and live events. Its audiences can access these streams, as well as the station's core radio programming, worldwide for free.

Facebook, Twitter, text messaging and other social media form part of the universe of interactive possibilities in which listeners, both domestic and foreign, participate with these stations. These interactions can and do include such exchanges as indirect communication with friends and family through such traditional means as song requests, direct communication with on-air personalities, feedback on content, user comments on station-hosted events, domestic users contributing news items and foreign users reporting on having received the station's Internet stream. Staff at 103 FM estimated their Facebook followers at over thirty thousand and also cite local media statistics suggesting that the launch of its website was accompanied by an audience boost of some 18 per cent in the sixteen-to-twenty-five age group.

Henderson (interview by author, 15 July 2015) of 103 FM indicated that the possibilities of digital broadcasting hold attraction beyond the now well-established Internet streaming technologies, noting that the potential for broadcasting multiple streams either on the air or via the Internet provides new and expanded opportunities: "So we could have about four channels simultaneously running different kinds of music. So we could play religious Indian music in one channel, Bollywood now – whatever the market wants from us, we could put there."

Yet, despite the technical and business potentials involved, the Trinidad Indian radio media sector remains bound to relatively traditional conceptions of mass media and advertising opportunities. As Henderson (ibid.) explained:

> As a viable business proposition, the reason why we are treading lightly or slowly is because we have not yet been able to really convince the local business community that it is a good proposition to advertise on the Internet. Many of the people in the business community are still very traditional. So we have to play along with that, you know? And whatever little advertising we get now, well, we're grateful for it, but it's still the traditional broadcast.

Listener and Community Engagement

One of the distinguishing characteristics of Indian radio stations in Trinidad has been their efforts at community engagement through listener involvement. Managers and entrepreneurs in this media sectors recount that early listeners and members of the target community were known to come forward and offer their record collections, their sponsorship and even their financial resources. The dimensions of involvement over their years of existence have been diverse. In the beginning of the Indian-radio era, the primary form of listener engagement was by way of listener calls to the stations to comment on programming or hosts, to request song dedications, to participate in competitions and to express greetings to individuals and communities. Soon, the spread and adoption of mobile and Internet technologies expanded the possibilities for interaction to include text messages and social media postings.

In pursuit of increased audience involvement, these stations conceived and staged several public events designed to draw their listeners and to build the station brands. Such events were branded variously as *mela* meaning "gathering" or "lime" (a festive gathering or hanging-out) or other similar terms. In some cases a competitive element might be included, such as the "curry duck" competition first made popular by 103 FM.

An important component of these stations' audience and community involvement has also come in the form of transnational linkages and events. As we have noted above, requests might be informally taped and exchanged, and early research indicated that stations experimented with simulcasts in which deejays in Trinidad and abroad hosted programmes while connected by phone.

Today, listener involvement includes multiple mediated modalities that facilitate both local and foreign listeners. Announcers routinely convey greeting to listeners in the United States and Canada from their families in Trinidad and in the opposite direction as well, while the listeners and their relatives in both places interact with the station and the announcers on Twitter and Facebook. Audience members' reactions to the Indian radio stations can therefore be gleaned in part from phone calls aired on the stations, Facebook and Twitter posts and from conversations with listeners.

During the course of fieldwork for the present project, numerous casual listeners in Trinidad were willing to comment on their engagement with the Indian radio stations and expressed a diverse set of responses. Many, in fact volunteered to share their opinions (none of the names used here are the real names of the participants).

A forty-eight-year-old Indo-Trinidadian female from Chase Village in central Trinidad whom we shall call Sharmilla, for example, indicated that she listened to all of the stations in random rotation at different times, and pinned her affinity for the Indian radio stations in Trinidad to her broader involvement in and exposure to Indian music in her upbringing: "My background was based on Indian music. My father was a singer, [and] played drums. Our whole family enjoyed Indian music. My brother used to be in an orchestra." Ashok, a young male listener who agreed to be interviewed, also made the connection between the music on Indian radio and his own musical experience and culture, saying that while he enjoyed the more modern and upbeat music on Indian radio, it also related to his academic study and practice of music and gave him "a little exposure to music of my own culture".

Sharmilla said that since the introduction of Indian radio stations, "We all feel more lively." This is a reference not just to her family but also the community of Indo-Trinidadians as she perceived it. She felt that the stations are a source of information as well as entertainment and that they stimulate discussions and debate with friends and family, admitting, for example, that she got most of her news from these stations. Sharmilla, like many other listeners, indicated that she calls the stations to send song requests and greetings to her family members both locally and abroad. She said she frequently calls and talks with several of the announcers on different stations and has developed social relationships with a number of them even as they have shifted from one station or another.

Carol, another female Indo-Trinidadian listener, was fifty at the time of our interview. She lived in the Tacarigua area, in a predominantly Afro-Trinidadian neighbourhood, and indicated that she enjoyed the educational programming on the Indian radio stations, the musical programming and some specific programmes that she would tune in to at particular times of the day on particular stations. She appreciates discovering new music and songs that she has never heard before but also listens for the older songs – or, as she describes them, "the ones which I long to hear often". Herself a Hindu, she perceived the stations to be good sources of information, providing most of her news and also information about traditional religions "including Islam and Hinduism". Several other interviewees also expressed this sentiment and it may be that the particular information they reference here has not traditionally had outlets outside of the formal religious setting such as the mosque or temple, except for brief items on broadcast media, such as the programme that was known as *"Meditation"* on Trinidad and Tobago Television for many years (though it should be noted that in more recent times, the variety of religious programming has expanded tremendously with the introduction of cable television and dedicated channels for religious programming, including the TV station from Radio Jaagriti). Several interviewees associated their listening to Indian radio with feelings of increased religiosity or awareness of their religious identity.

Carol said that most of her friends and family members were not avid listeners and she did not call the stations for requests, as many other listeners commonly do. Asked how she felt about Indian radio in Trinidad, Carol replied, "Well, it puts a sort of happiness in me." Other listeners similarly said that the stations provide "peace of mind". This emotional connection to the Hindi music and content perceived to be connected to Indian heritage was also evident in the comments of other interviewees. Notably, this was the case even when (as with the vast majority of listeners) the listener has no grasp of either Hindi or its local variants. As one interviewee put it, "I get great pleasure – although I don't understand – I get great pleasure from it. It keeps me going all the time." Another noted: "Sometimes you get a peace of mind. You know, even if you don't understand the songs, you get a lot out of just knowing that there is a lot of meaning in it." Many others used terms such as "support" and "comfort" to describe the overall effect of these stations on their lives. Asked if Indian radio stations had affected her life in any way, one listener described a calming effect in the following terms: "Oh, yes . . . it calms me down a lot. It might have me in such a good mood that I might forget to argue with the children. It takes away a lot of my tension." For some older interviewees, there was reference to feelings of nostalgia, since many of the songs played include old standards from a time when the range of songs available was not that wide. One collaborator noted that he most liked "the old songs that are being played", because he "grew up on those old songs". Another informant offered, "I get a lot of memories of my youth and a lot of inspiration towards the future."

Leela, an eighteen-year-old Indo-Trinidadian, listened only for music, but had her own particular favourite because she felt that many of the others did not do a very good job of programming and presenting (others also indicated that they have gravitated to one particular station over the years). She felt that the Indian radio stations provided useful information, but not for news and current affairs, saying, instead, "You listen and you learn about heritage . . . culture."

Leela also called these stations and remembered chatting with some of the on-air personalities, but her reasons for calling the stations were not so much for the purpose of making dedications, but more frequently to participate in the many competitions that the stations present. While calling up the stations for one reason or another (and, more recently exchanging texts and social media messages) might be a hallmark of the audience engagement, other audience members indicated that even though they do not call or interact with the announcers and personalities, they perceive themselves as having a kind of para-social interaction (Schiappa, Gregg and Hewes 2005) with the radio staff. Interestingly, particularly considering her relative youth, Leela said that she did not necessarily grow up with Indian music outside the context of the Indian radio stations, but learned to appreciate this music from the stations. This was a sentiment echoed by several other interviewees of different ages who suggested that the presence and social roles of Indian radio served as a kind of magnet that drew them to an appreciation of Indian music and culture to which they might have been ambivalent before. Others did not credit the stations with awakening or galvanizing their cultural identity, but asserted that their affinity for these stations arose from their pre-existing cultural identity, or, as one person put it: "I enjoy it because it is my culture." Another noted: "The Indian stations, they have good songs. Even though I cannot understand the meaning of the songs, the rhythm is nice and it makes me feel, well, I am a Hindu and it fulfils me."

Geeta, a forty-year-old Indo-Trinidadian woman from the San Fernando area in South Trinidad, made an intergenerational connection, placing the Indian radio stations in the context of cultural transmission and identity: "Well, as a parent of young children, I would say, very emphatically, that it provides our children to know their culture, their music and their roots, because, somewhere in all of that, they will find something to fit their life. They will know where they come from and where they will go from here." Asha, forty-three, made a similar connection with reference to the past, saying that the Indian radio stations provide "a better understanding of our Indian language, our roots, our culture which our forefathers brought to this country".

Ashok, like Leela, indicated that he had a singular preference for one station. While several audience members indicated an attachment to one or another station, others suggested that they switch among the stations depending on what music or programming is on offer at particular times, but also because certain stations do a better job of certain kinds of programming. One avid listener to Indian radio, Vikram, demonstrated during a journey by car that he switched among the various stations on the basis of what he perceived as their particular strengths. Thus if he wanted to hear chutney music, he switched to his favoured station for that genre. If he wanted to listen to classic Hindi hits, that was another station. On the hour, he would switch to his favoured station for news, though he admitted that the newsreader who was on at the time was in the habit of butchering the English language.

Several of these informants described the Indian radio stations as having become ingrained in their social and cultural existence – suggesting that they provide a sense of belonging and that without them the community would now experience a serious void. One informant even suggested that without Indian radio in Trinidad, people in the community would go mad. Another described the affinity of the local community for this medium, arguing that people had become so accustomed to this kind of content that to lose it would be a disaster.

The scope of information-gathering through these stations has also experienced some morphology, with changing patterns of use and emerging media forms. While many of the interviewees in the present work talked about learning traditional, cultural or religious ideas from the Indian radio stations, they also indicated that these stations have often become their main sources of news and current-events information. As information sources, the stations have coupled their news broadcasts with social media and mobile technologies, providing listeners a diverse set of opportunities for both receiving information and sharing ideas. Thus listeners today might get their local news headlines on a live terrestrial broadcast, send a request for a song via mobile text message and also check on Bollywood news or announcer posts on a station's Facebook page.

Radio stations have cultivated these varied opportunities for interaction over time, in part to overcome the hesitation that many listeners feel over making live calls to request songs. Several of the stations have shifted in recent years from live on-air calls to either delayed requests or simply taking requests by text or social media posts. Much of this has had to do with technological and regulatory changes over the years that have seen the diffusion of mobile and broadband technologies in Trinidad and Tobago. Since 2004, several private carriers have been able to compete in the market for cell phones, as the government scrapped the monopoly of the state-run telecommunications agency. This change led to the introduction of low-cost mobile services, including pre-paid phone-card-based systems accessible to a wide range of users. These changes, along with increased competition for broadband Internet services and bundling with cable, have also made access via social media more readily available. At the same time, these stations have also pursued live events during which listeners have the chance to meet on-air personalities. One informant fondly recalled his experience at one such event, saying: "Sometimes, when they have the *Chutney Train*, the people from 103, I go to the *Chutney Train*. While the show is live, they will have the announcer, you can look at them, can actually talk to them – they are very nice people, as well."

As the stations have explored varied opportunities for audience interactions and reach, they have also explored a variety of informational content. For some listeners interviewed, the fact that many stations provide synopses or translations of songs (in the past this was done on air, while today these are equally likely to be posted on Facebook) serves as a kind of language instruction, or at least provides linguistic insights that were previously difficult to find. One informant noted: "I really like listening to the translation of Indian songs, it helps me learn, if just a little, then a little of the language. And you get a great feeling from it . . . as I learn about the meaning of the songs, I like to talk about that with my family."

Listeners also noted many other non-musical varieties of information, including horoscopes, recipes, religious instruction and Bollywood gossip. On the question of horoscopes, some older listeners remembered the initial horoscope offerings from an on-air pundit who used Indian astrological signs rather than the more common Western signs used elsewhere.

Another form of audience interaction that has reaped the benefits of new media and mobile technologies has been the audience competitions. Announcers will invite listeners to call, text, or post responses to questions or challenges. The range of these competitions has increased with the ability to post pictures and videos that are then sometimes judged by other audience members.

While most of the listeners interviewed in Trinidad did not use the audio or video streams on a regular basis, there is evidence from the Facebook posts of many audience members that these streams provide a substitute for local listening or interaction with the station. Combined with Facebook, these streams allow listeners abroad to contribute to the broadcasts with requests, and to engage in discussions about programming.

The presence of Indian radio stations on-air and online is also supplemented in the target community by a fair level of involvement in the community, both, as noted, through formal events, but also more casually through social presence, including everyday discussions among friends and family. Fareesha, a thirty-nine-year-old Indo-Trinidadian, was among those who described the Indian radio content as being a part of her social interactions, saying, "Sometimes there is a subject that you bring up in conversation that you bring up in work. You know, if that person was listening and you want to get their views on it. And my children would question me, well, 'Mommy, what is this, and what is that?'"

Examining Foreign Listeners

The increasing availability of both streaming technologies and social media has simultaneously created the potential for increased reach and the potential for increased interaction with even (traditional) local media such as radio. Audience interactions with Trinidad's radio stations occur today through a wide variety of media platforms, including phone calls, text messages, Twitter and others. However, the popularity of Facebook, particularly as a forum for maintaining transnational family ties among Indo-Trinidadians, has made this the preferred platform today for audiences to engage with Indian radio stations from Trinidad. This social media platform thus enables the examination of foreign and domestic listeners who engage with the stations and a comparison of their use of and responses to the content of these stations.

Prior research (Fursich and Robins 2002; Halavais 2000; S.N. Mohammed 2012b) has suggested a strong domestic focus in the use of global media (and global uses of local media). In such previous studies, scholars have identified a propensity for dispersed audiences to use globally connected media to reconstruct or reinforce identity notions that may be centred in other places and to construct notions of home while being displaced from such a home. Following on this thread of evaluating the global and local uses of modern media, Mohammed and Thombre (2017) investigated the user comments on the Facebook pages of six Indian radio stations from Trinidad whose audio feeds were available online. The authors coded a total of 568 posts, identifying the poster locations (where given) and the nature of the content, including such variables as whether they contained requests, contest engagements, comments on programming, expressions of nostalgia and several other items. Additionally, posts were evaluated for expressions of ethnic identity.

In that study, the authors found that (where location could be determined) domestic listeners (that is, those physically located in Trinidad and Tobago) accounted for 51.1 per cent of the cases analysed, while foreign posters accounted for 33.6 per cent, a difference greater than could be accounted for by chance. This alone would suggest that domestic listeners were more likely than foreign listeners to engage with the station content by commenting on Facebook. Posters from the United States accounted for 18.5 per cent of total posts and 7.6 per cent were from Canada, while the remaining foreign posts came from several locations, including the United Kingdom, Jamaica and Guyana.

When the authors corrected for the differing proportions of foreign and domestic posters, the study found that the two groups did not differ significantly in their propensity to send greetings to the announcers or in the likelihood of their making comments or inquiries about the station, though foreign listeners were significantly more likely to enquire about the audio streams. Some 30 per cent of all comments included a greeting or request component and (correcting for the different proportions of foreign and domestic posters) both groups were equally likely to engage in this particular use of the Facebook pages of Trinidad Indian radio stations. Foreign listeners, however, were significantly more likely to include identity or nostalgia markers in their comments and to display ethnic markers, while domestic listeners were more likely to comment on local news and engage in competitions.

This snapshot of the foreign listeners to Trinidad's Indian radio station offers just the slightest insight into the global/local processes. The dynamics of assimilation and cultural maintenance involved in the foreign, diasporic listener's consumption of media from "back home" are quite different from the processes involved in the domestic listener's consumption of the same content. In the case of Trinidad Indian radio, however, both domestic and foreign listeners are subject to parallel processes of assimilation and cultural maintenance. Notably, while the foreign listeners made overt associations between their migrant identities and the content of the stations, domestic listeners made no measurable mention of such identities, suggesting, the study argued, that the sense of being migratory or dislocated has already faded in the "primary" or "old" diaspora situated in Trinidad and Tobago.

Conclusion

Examination of the global dimensions of Trinidad and Tobago's Indian-format radio stations reveals a complex set of interrelated communication processes. Well beyond the traditional notion of a radio-station broadcast, these stations now engage global audiences (often involving transnational families) in ongoing conversations fuelled by social media and often involving real-world interactions in special events at home and abroad. These complex interchanges require at the very least a rethinking of the orthodox notions of ethnic media that are prevalent in the communications literature. They also demand a broader look at issues of community and identity as virtual cultural enclaves emerge around these stations with little regard for their physical locations.

Contextualizing Indian Radio

This concluding chapter examines the overall findings of the present research in terms of broader ideas about cultural identity, historical globalisms and hybridity. It also considers some of the more contentious issues arising out of the research in which sources presented conflicting views on aspects of the Indian-format radio sector in Trinidad and Tobago. Among those issues is chutney, which is examined here not only for its relevance to radio, but also for its many connections to questions of hybridity, authenticity and identity.

Indian-format radio in Trinidad and Tobago, beyond its commercial success, raises numerous issues related to culture and identity. These issues are fore-grounded in Trinidad and Tobago because it is a relatively newly independent nation and one in which several ethnic groups with differing cultural outlooks compete for social space. Guilbault (2000, 435) argued that music in its many forms and extensions has always been part of what Michel Foucault has termed the "regime of truth" in West Indian politics, with "musical discourses" that "have not only reflected but effectively helped produce, sustain, and challenge relations of power". Much of the content of Indian radio is "Indian music" in any of several forms (from Bollywood film music to local chutney). The politics of music in Trinidad and Tobago features ongoing jostling for cultural space among many competing forms and traditions, including those that exercise the hegemony of national identity established through colonial avenues of power and influence such as calypso and steelpan, and others that have fought to enter that hegemonic space, such as chutney or its hybrid, chutney soca.

In this jostling for space there is evidence of change, even as chutney and chutney soca proliferate and draw large audiences with the help of Indian-format radio. Yet there is also evidence of the persistence of barriers. As *Caribbean Beat* magazine ("New Wave of Trinidad Music", January 2006) has noted: "Non-Indian radio stations rarely play chutney music and chutney events are patronised by a clientele that is still predominantly Indian."

There exist many ironies in the battles for cultural space. Not the least of these is the struggle to occupy a rightful place in national identity while also claiming some inherited tradition from elsewhere. Questions of legitimacy in the national sphere and authenticity in the ethnic and cultural spheres often hang uneasily in the cloud of claims and counter-claims or even fade against the emerging irrelevance of traditional modes of culture.

This jostling for position is not limited to external struggles, as many internal struggles also typify the "Indian music" genres. Manuel (2000a, 203) has argued that these politics of music bring into focus both the active nature of modern identity and its interplay with the traditional in the Caribbean:

> Modernity is the way in which individuals have become increasingly free – indeed, condemned – to choose their sense of identity self-consciously, rather than inherit it unquestioningly as a preordained given. The obligation to choose becomes particularly acute in a diasporic situation marked by a declining traditional cultural core and the presence of new cultural options and pressures. In Indo-Caribbean society, musical tastes serve as remarkably clear indices of the variety of forms of Indianness that different individuals currently cultivate. Some Indo-Caribbeans, including many film-music lovers, look to India in their quest for authentic Indianness, in some cases accepting the notion occasionally intimated from Indian nationals that Indo-Caribbean culture is inherently hybridized and impure.

Within the context of Indian-format radio, listeners and programmers have demonstrated that the audiences consume a wide variety of content. With musical choices ranging from local classical to chutney, nostalgic Bollywood hits and reggae remixes (to name only a few), audiences choose their favourite genres and content, switching when needed to maximize their listening pleasure. As Ramdeen and others noted, there is no restriction on these choices and Indo-Trinidadian audiences may listen to any, all or none of these, and may switch to an urban or adult contemporary station as their mood dictates. Yet several of the people interviewed here and numerous of the posts on social media platforms indicate something of the identity processes involved in their listening to Indian-format radio. The notion of peace of mind, hints of nostalgia for a time gone by, or a sense of ownership and involvement in the community created by these stations indicate that they provide some level of identity support.

Interviewees suggested that both the historical context of limited media opportunities for content that was focused on Indo-Trinidadian likes and concerns and the traditional exclusion of such content from the cultural mainstream meant that the arrival of Indian-format stations was seen as a welcome and satisfying development, providing a sense of validation and belonging. Notably, since engagement with these media has evolved from passive listening into social media interactions and event attendance, the notion of identity is not limited to personal choices, such as what music one chooses, but also linked to communal activities and discussions.

At the same time, not everyone who is a member of the Indo-Trinidadian ethnic group is also a fan of Indian-format radio. Several people interviewed in the current research did not habitually actively listen to these stations, but reported that they were, at most, passive listeners while others in their homes were tuned in. Other Indo-Trinidadians indicated that they did not listen to these stations at all. Well before the emergence of 103 FM, several decades of young Indo-Trinidadians had grown up with increasing levels of integration into the broader cultural mainstream. While most were exposed to Indian or local Hindi or Bhojpuri music, there were increasingly weak identity claims or connections to be made. Indeed, among many young Indo-Trinidadians in the 1970s, 1980s and 1990s, there was an active process of adoption of Western musical content, including not only country and western, but, increasingly, heavy metal and rap from the United States, as well as reggae and dancehall from Jamaica. As global media forces introduced numerous and varied options, Indo-Trinidadians could combine the influences of the mainstream creole folklore, those of the Indo-Trinidadian cultural heritage and numerous adaptations and inculcations of regional and foreign cultural elements of dress, music and social expectations. As with most cultural dynamics, one expects that force is exerted in both directions. Against the backdrop of what might be seen as this drift from traditions and established cultural norms, the Indian-format radio stations have created a counter-force that makes Indian-oriented content more accessible and familiar.

At the same time, these stations also encourage more complex processes of cultural hybridity. For example, on these stations the Indo-Trinidadian cultural past, characterized in part by prior generations' affinity for now-unintelligible Bollywood songs, is now redefined with the presentation of hybrid forms such as chutney, which includes English lyrics and resounds with the influences of Carnival and soca (now culturally very near and familiar to modern generations of Indo-Trinidadians). In popularizing these and other hybrids (such as reggae and soca remixes of Bollywood songs) and in perpetuating the entrenched globalisms evident in their promotion of Bollywood hits both nostalgic and new, Indian-format stations are providing an expanded array of cultural options to Indo-Trinidadians who listen to them. Out of these options emerge new configurations of identity, existing not as generalizable stereotypes of a group, but as individualized self-conceptions that include complex interrelationships among historical narratives, modern selves and notions of belonging to cultural subgroupings as well as to the larger nation.

Treason, National Cohesion and Business

An overriding concern and persistent criticism of the entire concept of Indian radio in Trinidad and Tobago is the argument that this ethnic media form presents a threat to national cohesion and represents at least the desire to remain apart from the mainstream of Trinidad and Tobago society. The implicit assumptions of this argument exclude the inherited Indian cultural traditions from social legitimacy and suggest that an integrated mainstream can only be achieved in terms of what Wilson (2012) has termed the creole cultural forms. Even while acknowledging (as with all modern cultural forms) the imaginary, pragmatic and syncretic nature of Indo-Trinidadian cultural expressions, the current research has revealed pointed indicators of the presumption of Indo-Trinidadian cultural forms as foreign or illegitimate. None of these indicators were more pointed than the reaction of civil servants and others at the mention of the start of the first Indian radio station. Dik Henderson even noted that the very idea was dismissed as "treasonous", though he excused the reaction on the grounds of ignorance:

> That only came about because by and large, at that time the people I went to deal with at all these public – what do you call it – civil-service offices were people who were town people. They didn't understand the Indian community at all. And they were – I

wouldn't say they were racist, but I would say they were just ignorant. I wouldn't even say they were discriminatory, just ignorant of the possibilities, right? Among some of my good friends too, personal friends, they couldn't understand: "What are you doing with an Indian station? That's not your background." So I said, "Well, it's business."

Though in the reference to "town people" Henderson revealed something of the traditional distinction between the Indo-Trinidadian as rural inhabitant and the Afro-Trinidadian and European as urban dwellers, he also offered that this kind of negative reaction from the "town people" has faded as the media sector has gained wide acceptance both in its target community and in the broader national environment. He argued that this wide acceptance and the imperatives of competition in business explain why additional Indian-oriented stations have emerged over the years since the inception of 103 FM.

Ramdeen (interview, 16 July 2015) agreed that the presence of the Indian stations has evolved over the years into greater acceptance, but noted that the reactions to ethnic media can be somewhat more complex than one might expect:

Even as recently as five, six years ago, when we proposed Indian TV, you know, we got a negative reaction, and we got enough negative reactions from even Indian people who control marketing for various companies. And we were like, "We just don't get this response." I think we've grown a long way since then. But to this day there are still two or three companies that don't advertise with us. And we have been given the clear impression that it is because we are an Indian or an Indian niche station.

But we are not that, you know, we are more than that. At the end of the day we don't see ourselves as an Indian station, we see ourselves as a truly national station. The fact is, some stations play calypso alone – so are they not national, then?

Sunil Ramdeen here approached one of the more contentious issues in the discussion of media in Trinidad and Tobago and one of the perennially unresolved polemics in Trinidad and Tobago society more broadly. While the charge is made that Indian radio, since it relies heavily on a distant and, for most, linguistically foreign heritage, is somehow anathema to an emerging national identity, the same charge is not made with regard to other components of the national cultural scheme. This is in part, perhaps, because the received and accepted notions of national identity emerge out of colonial narratives of nationhood that privileged race and ethnicity as markers of belonging – a glorified tribalism of its own (J. Singh 1996). Ironically, the same colonial schemata that produced these transplanted colonies and reinforced essential social divisions based on race are active in the post-colonial definition of the nations that have sought to unyoke themselves from colonial rule. The challenges of such emerging nationhood, involving disparate groups with differing cultural traditions and the negotiated cultural spaces that emerge, are indeed complex.

In this syncretized nation, even the labels of "Indian radio" or "Indian niche radio" inevitably become fluid and subject to the forces surrounding it. Ramdeen (interview, 16 July 2015) contended, for example, that it is impossible to separate these influences, pointing out that despite the (perhaps necessary) mixture of music played, his station chiefly serves to expose and promote talented locals from all genres: "So you can't have a country as mixed as ours and not play a significant amount of music. We released ourselves as a truly national station because we play soca, we play Indian music, we play chutney, we play dancehall, we play reggae, we play pop music. But you know, at the end of the day we support local artists more than anything else."

We have already examined the overt reactions from competing mainstream stations with small amounts of Indian-oriented content against 103 FM when the dedicated Indian-content station came on the air and the launch of competing stations within a few months of 103's debut. Today, the fight for scarce resources in a heavily fragmented media market drives intense business competition among the various shades of Indo-Trinidadian stations, with each attempting to jostle for a particular market niche.

This competition for market share is primarily a business process in which each station attempts to garner the largest audience in its particular scope of programming. As in other markets, stations have varied their content and focus to suit particular audiences, whether that be religious listeners, nostalgic Hindi film-song fans or chutney enthusiasts. To some extent, there is also station-jumping based on time of day and context, so that listeners may switch among those stations offering party music for weekend activities and family gatherings, while they may listen to stations carrying religious programming during seasonal observances. Interviewees almost uniformly described such switching activity, based on these and other factors such as seeking their favourite on-air personalities.

The entanglement of race and politics in Trinidad and Tobago inevitably also means that beyond questions of national cohesion and identity, more pragmatic questions of political affiliations and influences surround Indian radio. Several commentators have pointed, for example, to the possibility that Basdeo Panday's

United National Congress's electoral victory in 1995 was at least in part due to the influence of 103 FM, which was launched only two years prior. Verma (2000, 353) quoted an unnamed 103 FM on-air personality saying:

> 103 FM brought those Indians together who would never have met before in their lives. And when people meet, it is a strength for their community, because ideas flow. They share ideas and the idea expands and travels. It caused a reawakening, and people began to realize that 103 FM was beginning to bring Indians together as never before … and in a real threatening way, from the political perspective.

Verma (ibid., 353–54) argued that "where political organizational efforts, as well as the print media had failed, the radio stations succeeded … through the radio stations, the 'pan-Indianness' at the grass-root level that had historically been sought but never quite attained, was realized, at least at the cultural level" and with political impact: "Programmes such as talk shows and call-in sessions focusing around contemporary national social or political issues were avenues through which an 'Indo-Trinidadian' position from the grassroots was given voice. The mutual exchange of views and opinions of Indo-Trinidadians in a 'home environment' further consolidated their sense of oneness against the Afro-Trinidadian 'other'."

To be clear, several other stations with overtly political (and often deliberately provocative) content have emerged in the Trinidad and Tobago media market over the years as well. Stations such as Power 102 FM, Talk City 91.1 FM and i95.5 FM are known for politically minded and outspoken figures such as Ricardo "Gladiator" Welch, who is a self-described PNM party activist ("'Gladiator' Plans to Sue Rowley, PNM", *Newsday*, 12 April 2015, 1), and long-time PNM supporter and financier George "Umbala" Joseph. Yet, despite the political implications of Indian radio, Henderson insisted that 103 FM remains politically neutral, arguing:

> We have never entertained any political bias. But people know that we are in the Indian community, we have an audience in the community, and if they want to get in the Indian community they advertise with us. The PNM, when they come, they're ready, will advertise with us. And they have in the past, during election time.
>
> But I don't think we've ever had a problem with this. We have always been firm that we are non-political. Right now there are a couple of songs out, I've been told that they're being produced by the UNC [United National Congress] and the PNM, and we were just saying this morning that we won't play them. We want to stay neutral.

But we are in the business of advertising, so yes, we broadcast the public meetings. And we take their advertising . . . but we are non-political.

In this competition for market share, the business competition also devolves into more complex contestations over authenticity and relevance. Officials at the dedicated Hindu religious station pointed to the fact that they distinguished themselves by eschewing party, liquor and tobacco advertising. Officials at other stations also took issue with their competitors' claims and practices in the industry subsector. At WIN 101 FM, general manager Sunil Ramdeen challenged the seemingly exclusivist claims of other stations which attempted to provide solely Indian content, arguing that as a national station WIN 101 FM attempted to present a mix of reggae, soca, Indian and other music, in keeping with the varying cultural currents in Trinidad and Tobago and its regional environment. An announcer at a competing station, who spoke on condition of anonymity, suggested that such a mixed approach was absurd, since the mood and context of content such as classic Hindi film music was damaged by the intrusion of soca and other musical forms that he considered somewhat coarser (strongly suggesting a privileged position for Indian film music).

Officials and announcers in the industry even voiced some criticism of 103 FM's claims to being the first Indian radio station. Indeed, one person with the TBC group continued to argue that Radio Trinidad was in fact the first Indian radio station in Trinidad and Tobago, because it broadcast the first Indian programming in 1947. One announcer at a competitor even suggested that 103 FM could no longer claim to be an Indian station, because its concessions to local chutney music and its involvement with parties and concert events marked it as a party station instead.

Notions of Indian-format radio being treasonous, divisive or presenting threats to national cohesion are also frequently expressed by detractors in letters to the editor and in call-in programmes. Detractors characterize the focus on Indian cultural expressions as not just insular, but also as unfair to the Afro-Trinidadian population, as no African-oriented content exists on the local airwaves. Primarily, however, the notion of an Indian focus being anathema to the national interest hails back to early suspicions of indentured populations, and also assumes that Indo-Trinidadians as a group are somehow lesser members of the nation by nature of their cultural background or by their expression of it. Yet the Indian-format stations themselves, by evidence of their content and the statements of their officials, do not have fixed notions of cultural apartness.

Their very approaches suggest hybridity in its many forms, while their diverse ownership and management structures also suggest that, by and large, while they may focus on a culturally specific audience, their aim is to reach the broadest possible national audience.

The Chutney "Problem", Indian-Format Radio and Hybrid Identities

An intricate web of interrelationships exists between the Indian-format radio sector and the evolution and support of the musical form commonly known as "chutney". Many of the participants in the present fieldwork interviews (both station officials and listeners) spontaneously raised the issue of chutney, not just as a musical form but also as a kind of overall cultural expression. Chutney, therefore, is more than a kind of singing or music. It also connotes a particular kind of dance, as well as a challenge to traditional notions of sexual morality and expression (many of which evolved out of specious notions about Indian traditional sexuality shaped over decades by misinformation, seclusion and colonial pressures of prudish Victorian pretensions both in India and in the islands). As C. Gupta (2011, 13) has noted, the British "exercised moral surveillance and used apparently 'obscene', lascivious, and sexually promiscuous activities prevalent in Indian society to 'justify' their rule and impose in part certain mores of Victorian chastity".

Ramnarine (1996, 148) commented that "as a tradition which has developed in diasporic contexts, drawing upon diverse cultural and musical elements, chutney . . . is a reflection of the variety of identities adopted by individuals". Thus an analysis of Indian radio in Trinidad must deal with chutney. With similarities to other hybridized, syncretized and culturally differentiated subculture phenomena such as flamenco and hip-hop, chutney is expressed primarily in music and dance, but also connotes broader cultural elements, including linguistic peculiarities, dress and adornment and even attitudes and beliefs.

Ramnarine (ibid., 133) described chutney as "a contemporary Indian-Caribbean musical genre which displays influences from diverse sources", including "Indian folk traditions, devotional songs and film music" as well as calypso, soca and rap. Baksh (2014, 152) argued that chutney does more than simply express Indo-Trinidadian identity, but rather "interrogates and confronts a variety of interrelated struggles over multiple identities – musical, socio-cultural and racial – in a bid to reinscribe broader definitions of Trini-dadianness and challenge traditional notions of Indianness".

A simple hybrid notion of chutney often explains this cultural subcomplex as a creolization of traditional Indian forms – particularly as an Indo-Trinidadian adaptation of creole kaiso and soca forms. This oversimplification incorrectly reduces chutney to a derived facsimile of dominant Trinidad Carnival forms and reflects an ongoing tendency to interpret Indo-Trinidadian cultural iden-tity (however imagined or constructed) through the dominant creole lens. The interrelationships among chutney and soca (as noted in Ramnarine above) may be much more complex. In fact, Niranjana (2006, 88–89) (with the backing of famous fusion musician and sitarist Mungal Patasar) is among those who contend that soca itself emerged from local Indian music forms: "In the 1980s, a new form called soul-calypso, or soca, emerged, claimed by its inventor Lord Shorty (later Ras Shorty I) to have its inspiration in East Indian music. Shorty's songs 'Indrani' (1973) and 'Kelogee Bulbul' (1974) provided the genesis of the soca, marking clearly the Indian influence on calypso."

The orthodox history (note Ramnarine's reference above to "marriage cel-ebrations") suggests that chutney emerged out of the *matikor* and "cooking-night" song and dance traditions (Bergman 2008; Ramnarine 1996). Ramnarine (ibid., 151), in fact described chutney as "hitherto predominantly female trad-itions performed in private and rural contexts (such as *mathkor*), which have been transformed in to a popular genre enjoying current favour".

Indo-Trinidadian Hindu weddings traditionally take place over a week-end, including observances on Friday evening and food preparations (cooking night) on Saturday. The *matikor* or *mattee korwah* (dirt-digging) ceremony takes place on Friday evenings and involves a procession of women (including the bride-to-be) to a riverside or other clean running water, where traditional tassa drumming accompanies often-ribald dances and singing. The staging of formal live singing of traditional songs at the "cooking night" also cited as a source of chutney music is a faded tradition, and has largely been replaced for several decades in most weddings by loud festive recorded music and, in some instances, modern live bands with electronic instruments (Hardial Gosein, interview by author, 9 April 2016).

Mungal Patasar (1998, 69), while agreeing that women's wedding songs were an influence, has clearly drawn attention to a broader range of musical expres-sions that are relevant to the emergence of the modern chutney form:

During the period 1917 to 1940 women's songs took precedence. These songs were associated with functions like *chuthee* and *barahee*. *Chuthee* is celebrated six days after a child is born, while *barahee* is celebrated twelve days after birth. These celebrations gave women the opportunity to sing, dance and enjoy themselves. At weddings, there were several celebrations that allowed women to enjoy themselves without the restrictions of having the males around. There were songs instructing the young bride about sexual rites, and what to expect on the *suhaag raat* (consummation night). These were done in gay and humorous tones in the presence of the blushing bride.

It is very likely that all these cultural practices influenced the form and expressions of what is called chutney today. Alongside these relatively ortho-dox practices, however, were other changes in musical expressions (some quite radical) that may have been at least equal contributors to the emergence of chutney – particularly by removing Indo-Trinidadian music from its bonds to tradition and religion as in wedding ceremonies (or in much more complicated interweavings during Muharram Shi'a or Hosay observances).

To find these, we must move out of traditional domains and examine the commercial success of several local and regional singers and musical groups who transformed the musical legacies received from traditional folkways, com-mercial Bollywood, and what was often called "local classical" (or just "classi-cal") music. As Ramnarine (2002) has noted, the designation of "local classical" includes what Manuel (2000a) and others have called *tan* singing, particularly since the latter term is rarely, if ever, used among Indo-Trinidadians. Indeed, Manuel (ibid.) has noted that many of the academic typologies of tradition and formal elements of music theory were often lost on the musicians he inter-viewed. Some Indo-Trinidadians do have a common understanding of what they call "tent" singing, which referred primarily to the music and songs ren-dered during the cooking-night festivities, which were Saturday-night obser-vances at local Hindu weddings while food was prepared for the following day. Some of those interviewed identified this "tent singing" as "local classical".

Manuel's (ibid.) research recorded complaints from the singers of local clas-sical music that more contemporary forms such as chutney and Bollywood film music were eclipsing their traditional music. Consistent with this perspec-tive, veteran musician Hardial Gosein (interview, 9 April 2016) recounted the gradual erosion of the "tent singing" form:

What used to happen, when you go out there on a cooking night, they used to have classical singing. They called it "tent singing". You hardly had bands really playing band music, right? So you would get them classical singers singing. You would get about five or six of them singing classical, until [in] the morning, you find it was only a set of older people, the older folks, sitting down listening to the songs. You hardly could find young people sit down and listening to those songs, you know.

In several variants, this local classical music, with only limited availability on records, was popular at small gatherings, sometimes recorded on tape and exchanged among enthusiasts during the 1970s and 1980s on cassette. Less frequently, recordings were arranged on an ad hoc basis and pressed into vinyl records – usually singles. It was from this "local classical" tradition that several singers emerged who produced a music that was more open to up-tempo beats and included some mix of English with Bhojpuri lyrics – some even being primarily in Indo-Trinidadian English. Blogger ChutneyRoots (2009) credited the emergence of chutney to stage shows hosted by the Mohammed brothers, starting in the 1960s:

> One of their protégés, Sundar Popo, who is now considered a legend in the pioneer [sic] of chutney music, enjoyed great popularity from the 1970s onwards with his light chutney compositions, which included Hindi and English verses termed "local song". But more than a decade would pass before the definitive public emergence of chutney would take place, and quite dramatically, in the mid 1980s. Weekend chutney dance-fetes hosted by the Mohammeds became for the first time, popular on a mass level, regularly attracting hundreds of patrons.

Sundar Popo's numerous commercial success, including "Nana and Nanee" and "Scorpion Gyul", enjoyed copious record and cassette sales over several decades, and sometimes gained popularity through these recordings being played at family events, community fairs (bazaars) and similar events (often being considered too risqué for radio airplay).

During the 1970s, chutney as an emergent form faced competition from more established local classical music, with the novel form perceived as having less of a connection to Indian roots and more in common with creole influences, particularly with its use of English lyrics in part or whole. This was so even though by that time, the lingua franca among the Indo-Trinidadian community was clearly English and the majority of audiences did not understand the hybridized Hindi/Bhojpuri forms used in local classical songs. The upbeat compositions of what would become the chutney genre also lent themselves to being performed with the backing of running bass lines and brass accompaniments associated with kaiso and soca forms (a comparison of Sundar Popo's original recordings with later ones and various remixes reveals the addition of soca-like bass lines and kaiso brass backing in many later versions). Several collaborators commented that at the emergence of this musical style, suggestive lyrics notwithstanding, the upbeat rhythms were also treated with some suspicion.

Beyond the region, ChutneyRoots (2009) has also dated the spread of chutney to the metropolitan cities in the early 1980s, and demonstrated Sundar Popo's continuing influence: "By 1982, chutney music began to cross the ocean from Trinidad and other Caribbean countries to the United States of America (USA). In the same year, Sundar Popo and chutney singer Drupatie Ramgoonai thrilled audiences of some 2,000 at Madison Square Garden's Felt Forum. The show was hosted by Mohan Jaikaran, who at that time was chief executive officer of Jamaica Me Crazy Records." Jaikaran would later go on to be a pioneer of Indian radio in Trinidad and was associated with the WIN 101/Masala radio and television organization until his death in 2015. He would also be mentioned when *Billboard* magazine (Fergusson 1996, 90) noted the increasing commercial importance and cultural and political significance of chutney in Trinidad and abroad.

The associations of chutney with subversion of traditional culture and with cultural resistance to powerful social forces, coupled with its reputation for using sexuality, humour and satire, suggest the possibility of invoking Bakhtin's (1968) Rabelaisian Carnival. Here it is important to emphasize that this notion of the grotesque in Rabelais differs somewhat from its usual connotation of something that is repulsive or twisted. While Rabelais' grotesque certainly includes that conventional sense, it also involves, for Bakhtin, the sense of involvement and fascination particularly with the symbolic destruction of repressive moral and social codes clearly evident in chutney's vast audiences within Trinidad, Guyana and venues further afield. The Rabelaisian emphasis on the body as a tool of protest finds parallel in chutney's blatant and provocative use of sexuality to challenge the often implicit boundaries of gender and (imaginary-constructed) sexual moral conservatisms traditionally imposed within the Indo-Trinidadian community. Relevant here, as well, is Bakhtin's emphasis on laughter as both a tool of involvement and a mechanism of protest, since much of chutney includes humour alongside sexual and political challenges to tradition and power.

Complicating this analysis is the possibility of also associating Trinidad's better-known mainstream pre-Lenten carnival with Rabelais, an approach alluded to in Alleyne-Dettmers (2002, 243), which described Carnival "as a ritual site that can and does facilitate a unique forum to dissect the social order and to carve out a public power space for political yet cultural expression of dissident groups". Though more clearly evident in the older forms of this Carnival, grotesque forms persist in modern observances through ole mas and J'Ouvert, which stand as stark counterpoint to the colourful and largely conformist displays of expensive costumes observed in the commercially driven parades. Thus from a broader perspective of Trinidad's cultural evolution, chutney can be likened to (and perhaps is not completely separate from) the creole traditions of protest and subaltern resistance embodied in mas, which originally emphasized visual parodies of power and authority (Green and Scher 2007; Mason 1998), and kaiso, still marked by its emphasis on social commentary (Rohlehr 1998; Warner 1982). This may account (at least in part) for some of the implicit resistance to chutney and the criticism that it adopts the negative elements of creole tradition.

The relationship between chutney in its various forms (that is, primarily recorded music, live concerts and party events) and Trinidad's Indian radio stations has not been a simple one. Initially, chutney music provided an opportunity for Indian radio stations to broadcast music of local origination that was easy to source, popular and relatively inexpensive. This claim to indigenousness, along with the popularity of chutney and its ability to draw audiences both to radio broadcasts and live community parties, propelled what appeared to be an obvious symbiosis between the chutney artistes and producers on the one hand, and Indian radio-station operators on the other. Chutney performers received airplay and promotion for their events, while radio stations received valuable audience share and listener involvement. The draw of chutney singers has frequently been the main force behind sold-out live radio-station-sponsored events in Trinidad and abroad.

Despite this relatively obvious and productive symbiosis, however, several issues emerged that have made this relationship between chutney and Indian radio stations somewhat more complicated and notably more fragile. Several informants involved in Indian radio broadcasting in Trinidad indicated that their well-established involvement with chutney had been something of a mixed blessing, and admitted to feeling somewhat let down and betrayed after devoting much of their programming and promotions to chutney for many years. More than one participant noted that the quality of the chutney offerings had fallen away precipitously over the years, with poor and offensive lyrics and poor musical accompaniment becoming the norm. Hemant Saith (interview, 15 July 2015) at 103 FM, for example, cited chutney music as a persistent challenge, saying:

> People still hate chutney to a point, because it hasn't really improved – if anything, it has gotten worse, where we have a lot more people now wanting to do chutney and they don't know the first things about doing music. So the music is weak, the lyrics are weak, which is why we had to pull back trying to support the local artists; it backfired on us. Because they weren't – we were giving them the radio station – but they weren't giving back in terms of quality. I mean, you can do what you want, people are going to listen and enjoy what they want, and they weren't enjoying it and they told us no. And it would have affected the station audience-wise and also business-wise if we didn't pull back.

Such a reaction does not emerge without the broader social context. A complex set of norms and expectations have evolved in Indo-Trinidadian society around notions of morality and propriety. These are partly traditional. However, they are also, in large part, imagined and cultivated out of distant (and often conflicting) oral traditions from differing ancestral groups who hailed from culturally diverse areas of India, muddled through the passage of time and oral retellings in lieu of literacy, cobbled together in the light of incompatibilities of regional traditions, pressures of the creole experience and emerging norms of Western religious and social standards of behaviour. These imaginary codes are enforced through pressures from various groups who claim to be authorities on morality and Indian culture, and from among what Johnson (2009, 131) calls the "Hindu bourgeoisie class". Notably, without the formal structures of caste or class that would empower such pressures, condemnation of chutney emerges from such organizations as the Hindu Women's Organization and the Sanatan Dharma Maha Sabha. Niranjana (2000, 128) described diverse responses to the cultural form in the following terms: "'Chutney is breaking up homes and bringing disgrace,' proclaimed a letter-writer in the *Sunday Express*. 'Culture means refinement, and this is not culture,' declared a participant in a seminar on the chutney phenomenon. The Hindu Women's

Organization, a small but vocal urban group, demanded that the police intervene at chutney performances and enforce the law against vulgarity and obscenity."

In this context, notions of sexually conservative morality as well as the politics of cultural expression and the ethnopolitics of national leadership converge uneasily in the polysemous arena of chutney and its relationship with Indian-format stations in Trinidad and Tobago. A reasonably typical reaction may be found in letters to the editor such as the following from Trinidad's *Newsday* from a reader who had viewed a live telecast of 103 FM's tenth anniversary celebrations:

> As it were what passed for chutney was nothing more than vulgarity, semi nudity, crude lyrics, and of course, the alcoholic overtones of "rum till I die". One notes that as a result of the ongoing efforts of WABC 103 FM to win audience listenership, the average chutney routine is now a scandalous wining exercise, laced with hip hop, dub, and soca. The themes were as pathetic as they were senseless. Prominent among the disappointments were Ms Nisha Baksh's semi-nude appearance on stage, which had her looking more like a stripper than hot chutney diva. . . . Indeed, people who took their children to the show to see Indian culture in action must surely have been disappointed. (Lystra Marajh, "Sacrifice of Indian Culture", *Newsday*, 16 July 2003, 5)

While the chutney fan is free to make his or her determination of whether the genre is appropriate, it is also clear that the negotiation of the place of chutney is itself demonstrative of the jostling for place that typifies the negotiation of identities both within a cultural community and with regard to its positioning against external cultures. This genre is also important because it demonstrates the inevitability of hybridity, both in its original form and in its morphing into chutney soca.

As Indian-format stations negotiate their relationships with chutney, they also negotiate their position within the array of notions that define Indo-Trinidadian cultural norms, partly reflecting and partly defining the identity claims of this group. While Indian-format radio listeners varied in their responses to chutney, radio station officials expressed some trepidation, treating it like a double-edged sword – capable of drawing listenership, but also capable of damaging a station's reputation. Questions of authenticity and hybridity were of less concern to station programmers than the balance between audience draw and the risks of alienating some listeners through the broadcast of content that might be considered to be in poor taste. At the same time, there was little direct questioning of the hybridities or the globalized historical influences inherent in this emerging cultural form. Indian-format radio listeners and programmers both recognized the popularity of chutney, with little question as to the provenance of its various historical and hybridized influences or whether it could or should be evaluated for authenticity.

Conclusion: Indian Radio and (Imagined) Communities

We return here to the notion of imagined communities and invented traditions as articulated in the work of scholars such as Anderson (1983), S.N. Mohammed (2011) and Sollors (1987). These prior works and a body of continuing investigation have argued that what were once viewed as fixed historical traditions may now be viewed in terms of continuous processes of invention or reinvention, involving eclectic and sometimes random choices among communities on how they define themselves.

One of the most important influences of globalized information technologies has been their ability to defy geography in the creation of communities. Global information flows enable a much broader notion of community in which information-sharing and other identity loci may be widely dispersed. The present and prior research suggests that through the use of multiple media strategies, including traditional broadcast signals, Internet streams (often including live video from studios or events), mobile connections and social media, listeners to Indian-format radio from Trinidad are able to engage in several levels of community, prompting and reinforcing notions of culture and identity. Transnational family links are kept alive through greetings and song requests, while parasocial interactions are fostered through contact with on-air personalities. Whether listeners are located in Trinidad and Tobago or in the secondary diaspora, they are able to engage with a like-minded community who share interests in particular musical styles but who also reference common events and some level of shared heritage. Such virtual or imagined communities can also be made real when stations hold events at home and abroad meant precisely to draw their listeners.

While the globally connected Internet and other digital networked technologies most clearly demonstrate these propensities, the role of information across long distances was also important before communication was both global and instantaneous. Thus Indian films, music and even news were important sources of identity information for Indo-Trinidadians long before the modern era. Such influences, in the absence of continuous contact with their ancestral homeland, often formed the basis upon which Indo-Trinidadians have imagined their community into being. From the adoption of cultural norms on the screen to their identification with Hindi film songs, which for generations have been incomprehensible to most of them, media from India provided an important component of their communal definition of what it meant to be Indian in Trinidad and Tobago.

These imaginations were not exclusively founded on Indian media. Religious and academic leaders (and later, as we have seen, average community members) among the Indo-Trinidadians travelled to India and returned to spread ideas; missionaries and Indian government officials visited sporadically to encourage particular ideals; individuals and corporations engaged (and continue to engage) in trade and business deals; scholarships and other academic exchanges also fostered the spread of Indian ideas among the West Indians.

Today, satellite and cable channels from India and Indian radio stations provide the latest news from Bollywood on the airwaves and on social media, further supporting the idiosyncratic construction of Indo-Trinidadian culture and community in which the foreign ancestral India is reinforced.

While these diverse influences are undeniably important and directly referenced in the discourse of Indo-Trinidadian informants, they are also never exclusive in their effect. The imagination of Indo-Trinidadian identity also includes the inevitable influence of what has been called the creole folklore. In the controversies surrounding chutney and chutney soca and in the debates over how much soca should be played on an Indian radio station, it becomes clear that the self-imaginations of this community necessarily also include non-Indian influences from Trinidad and Tobago and the broader Caribbean. While it may be somewhat facile simply to make this assertion, we may rely on evidence to this effect from such phenomena as the increasing levels of soca and reggae mixes of Indian music on the radio stations examined here (and in prior research) as well as the assertions of programmers like Ramdeen and Hanoomansingh, who clearly articulated that the Indo-Trinidadian identity necessarily includes diverse influences, including those from the broader Trinidad and Tobago and Caribbean cultural matrices. While some of these stations identify themselves as hybrid or "mix"-oriented (one carried the name "*masala*," or mixed, in a previous iteration), it is important to note that none of the so-called Indian stations actually programme themselves in Hindi (almost no one in the audience would understand) or any other Indian language (at least twenty-two main languages are mentioned in the Indian constitution, but far more than this number exist). For this reason, the mixed English/ Hindi offerings of the broadcasts mark all of these stations as hybrid or mixed, while the English/Bhojpuri chutney and the soca and reggae remixes further mark the dynamic, eclectic and constitutive cultural matrix of this imagined community.

The very existence of "Indian" radio stations in Trinidad and Tobago, a small Caribbean nation, is itself a product of cultural imagination and imagined community.

Among the practitioners within this media sector there existed discord over what should constitute an "Indian" radio station. Some felt, for example, that only stations playing old Hindi film songs were legitimately "Indian". That particular assertion is fraught with ironies, since Hindi films and Hindi film songs (or, for that matter, Hindi language) do not represent all of India. As we have noted previously, All-India Radio refused to broadcast such songs after Indian independence for these very reasons. This particular insistence on a certain specific media artefact (the one to which the community had the most exposure, for commercial reasons) as a cornerstone of cultural beliefs also demonstrates the extent to which communities can build themselves on the most specious grounds. There is no standard of cultural authenticity (itself a specious concept) to determine a proper source of cultural imagination.

There is thus some variety in these imaginations of community and culture reflected in the various Indian radio stations themselves. While 103 FM and other stations hold public events characterized by music, merriment and the consumption of alcohol, Radio Jaagriti, on the other hand, refuses even to include advertising for parties or alcohol. This disparity reveals that the community for whom Jaagriti programmes imagines itself in terms of a religious purity (whether authentic or not) to which the broader community does not subscribe. As we have seen before, this particular set of beliefs stems from practices rooted in ancient Bramhin caste privileges cobbled together with the help of missionaries during the nineteenth and early twentieth centuries to form an orthodoxy that may not have exact precedents in actual Indian religion.

These observations appear to raise the fundamental question of legitimacy (or what we have discussed before as "authenticity") around "Indian" radio in

Trinidad and Tobago as a cultural phenomenon. From a traditional cultural perspective, it may be possible to argue with some conviction that this media phenomenon is not in fact "Indian" in any authentic sense and that its claims to "Indianness" are unfounded. Other factors such as linguistic discontinuity and a lack of any real connection to the Indian subcontinent may further bolster such a case. Trinidad "Indian" radio is, from this view, not just inauthentic, but, arguably, not "Indian" in any real sense.

However, the interviewees, both audience members and industry insiders, identify something primal in the reactions to Indian radio from their audiences. Among the listeners who donated their record collections, or the business people who donated money to Indian radio stations, there is clearly an identity claim, a sense of self and an appeal to the familiar in this media form. What started off as (and continues to be) primarily a business venture has evolved into a major cultural phenomenon.

So how does a clearly inauthentic, commercial venture become the locus of identity claims for a group of people and a site of cultural contention for a nation? This happens precisely because there is no discrepancy between the authentic and the inauthentic. They are both, culturally speaking, whatever the subscribing community imagine them to be.

References

Adams-Parham, Angel. 2004. "Diaspora, Community and Communication: Internet Use in Transnational Haiti". *Global Networks* 4 (2): 199–217.

Advisory Committee on Wireless Telegraphy. 1913. *Report of the Committee Appointed by the Postmaster General to Consider and Report on the Merits of the Existing Systems of Long Distance Wireless Telegraphy*. London: His Majesty's Stationery Office.

Akbar, Mohammad. 1946. "Letter to Kamaluddin Mohammed", 11 Janaury. Kamaluddin Mohammed Papers. Correspondence, January 1946–October 1948. Vol. Box 1, Folder 5. Special Collections, West Indiana, Alma Jordan Library, University of the West Indies, St Augustine.

———. 1947. "Letter to Kamaluddin Mohammed", 30 August. Kamaluddin Mohammed Papers. Correspondence, January 1946–October 1948. Vol. Box 1, Folder 5. Special Collections, West Indiana, Alma Jordan Library, University of the West Indies, St Augustine.

Alleyne-Dettmers, Patricia Tamara. 2002. "Black Kings: Aesthetic Representation in Carnival in Trinidad and London". *Black Music Research Journal* 22 (2): 241–58.

Anderson, Benedict. 1983. *Imagined Communities: Reflections on the Origin and Spread of Nationalism*. London: Verso.

Anduaga, Aitor. 2009. *Wireless and Empire: Geopolitics, Radio Industry, and Ionosphere in the British Empire, 1918–1939*. Oxford: Oxford University Press.

Bakhtin, Mikhail. 1968. *Rabelais and His World*. Translated by Helene Iswolsky. Bloomington: Indiana University Press.

Baksh, Darrell Gerohn. 2014. "Jep Sting Radica with Rum and Roti: Trinidadian Social Dynamics in Chutney Music". *Popular Music and Society* 37 (2): 152–68. doi:10.1080/03007766.2012.737593.

Balliger, Robin. 2006. "Empire in the Present: Exploring the Indies through the Cultural Geography of the Commonwealth". *Anthropological Forum* 16 (3): 277–89.

Bayart, Jean-François. 2005. *The Illusion of Cultural Identity*. Rev. ed. Translated by Steven Rendall, Janet Roitman and Jonathan Derrick. London: C. Hurst.

Beckford, George. 1972. *Persistent Poverty: Underdevelopment in Plantation Economies of the Third World*. Oxford: Oxford University Press.

Benhabib, Seyla. 2002. *The Claims of Culture: Equality and Diversity in the Global Era*. Princeton: Princeton University Press.

Berg, Jerome S. 2013. *The Early Shortwave Stations: A Broadcasting History through 1945*. Jefferson, NC: McFarland.

Bergman, Sara. 2008. "Matikor, Chutney, Odissi and Bollywood: Gender Negotiations in Indo-Trinidadian Dance". *Caribbean Review of Gender Studies* 1 (2): 1–28.

Betaudier, Holly. 2010. "Holly Betaudier TTT Pioneer Remembers". *TTT Pioneers*. 1 July. http://tttpioneers.org/archives/224.

Bhabha, Homi. 1994. *The Location of Culture*. New York: Routledge.

Boomert, Arie. 2016. *The Indigenous Peoples of Trinidad and Tobago from the First Settlers until Today*. Leiden, The Netherlands: Sidestone Press.

Bowden, Sue, David Clayton, and Alvaro Pereira. 2012. "Extending Access to Information, Education and Entertainment in the Developing Economies: Broadcast Technologies in the British Empire, 1951 to 1962". *Journal of International Development* 20 (8): 922–45. doi:10.1002/jid.2809.

Bratu, Roxana. 2014. "Portrayals of Romanian Migrants in Ethnic Media from Italy". *Journal of Comparative Research in Anthropology and Sociology* 5 (2): 199–217.

Brereton, Bridget. 1979. *Race Relations in Colonial Trinidad, 1870–1900*. Cambridge: Cambridge University Press.

Brodsky, Ira. 2008. "How Reginald Fessenden put Wireless on the Right Technological Footing". IEEE GLOBECOM 2008: 2008 IEEE Global Telecommunications Conference, New Orleans, Louisiana. doi:10.1109/GLOCOM.2008.ECP.687.

Brown, Aggrey, and Roderick Sanatan. 1987. *Talking with Whom?* Kingston: Caribbean Institute of Mass Communication, University of the West Indies.

Bruner, Edward M. 2005. *Culture on Tour: Ethnographies of Travel*. Chicago: University of Chicago Press.

Burke, P. 2009. *Cultural Hybridity*. Cambridge: Polity.

Candan, Menderes, and Uwe Hunger. 2008. "Nation Building Online: A Case Study of Kurdish Migrants in Germany". *German Policy Studies* 4 (4): 125–53.

ChutneyRoots. 2009. "The History of Chutney Music in Trinidad and Tobago (Part 3)". 7 October. *Chutney Pulse: All About Chutney Music and Chutney Personalities* (blog). http://chutneyontheweb.blogspot.com/2009/10/history-of-chutney-music-in-trinidad_07.html.

CIA (Central Intelligence Agency). 2016. *The World Factbook*. CIA.gov. https://www.cia.gov/library/publications/the-world-factbook/geos/td.html.

Colman, Anthony. 2014. "HCU Report". Parliament of Trinidad and Tobago, Port of Spain. http://www.ttparliament.org/documents/2243.pdf.

CSO (Central Statistical Office). 2017. "Midyear Estimates 2016". 9 January. http://cso.gov.tt/data/?productID=31-Population-Mid-Year-Estimates.

Cuthbert, Marlene. 1977. "Mass Media in National Development: Governmental Perspectives in Jamaica and Guyana". *Caribbean Quarterly* 23 (4): 90–105.

———. 1990. "A Model for a Small Independent News Agency: CANA and CANA Radio". In *Mass Media in the Caribbean*, edited by Stuart H. Surlin and Walter C. Soderland, 403–14. New York: Gordon and Breach.

Das, Priya. 2010. "The United Sound of Brown". *India Currents*, June: 36.

Demas, William G. 1975. *Change and Renewal in the Caribbean*. Bridgetown: Cedar Press.

Diethrich, Gregory Michael. 2004. *"Living in Both Sides of the World": Music, Diaspora and Nation in Trinidad*. Urbana: University of Illinois Press.

Douillet, Catherine M. 2005. "A Contradictory Callaloo: Ethnic Divisions and Mixing in Trinidad". PhD dissertation, University of Iowa.

Dunn, Hopeton. 2004a. "The Politics of the Media in the English-Speaking Caribbean". In *Who Owns the Media? Global Trends and Local Resistance*, edited by Pradip N. Thomas and Zaharom Nain, 69–96. New York: Palgrave.

———. 2004b. "Struggle, Innovation and Cultural Resistance". *Intermedia* 32 (3): 6–9.

Edmonds, Ennis B., and Michelle A. Gonzalez. 2010. *Caribbean Religious History: An Introduction*. New York: New York University Press.

Eisenlohr, Patrick. 2006. *Little India: Diaspora, Time, and Ethnolinguistic Belonging in Hindu Mauritius*. Berkeley: University of California Press. doi:10.1525/j.ctt1ppkdj.

Fejes, Fred. 1986. *Imperialism, Media, and the Good Neighbor: New Deal Foreign Policy and the United States Shortwave Broadcasting to Latin America*. Norwood, NJ: Ablex.

Fergus, Claudius. 2008. "Why an Atlantic Slave Trade?" *Journal of Caribbean History* 42 (1): 1–21.

Fergusson, Isaac. 1996. "Copyrights, New Government Change Tune of Trinidad: Radio among Topics at Island Soca Confab". *Billboard*, 6 January, 90.

———. 1997. "Carnival's Controversies". *Billboard*, 29 March, 1, 14.

Figueira, Darius. 2003. *Simboonath Capildeo: Lion of the Legislative Council, Father of Hindu Nationalism in Trinidad and Tobago*. Lincoln, NE: IUniverse.

Franco, Pamela R. 2007. "The Invention of Traditional Mas and the Politics of Gender". In *Trinidad Carnival: The Cultural Politics of a Transnational Festival*, edited by Garth L. Green and Philip W. Scher, 25–47. Bloomington: Indiana University Press.

Funk, Ray, and Donald R. Hill. 2007. "Will Calypso Doom Rock and Roll? The US Calypso Craze of 1957". In *Trinidad Carnival: The Cultural Politics of a Transnational Festival*, edited by Garth L. Green and Philip W. Scher, 178–97. Bloomington: Indiana University Press.

Fursich, E., and M.B. Robins. 2002. "Africa.com: The Self-Representation of Sub-Saharan Nations on the World Wide Web". *Critical Studies in Media Communication* 19 (2): 190–211.

Gamble, W. 1866. *Trinidad, Historical and Descriptive: Being a Narrative of Nine Years' Residence in the Island*. London: Yates and Alexander.

Garratt, Gerald Reginald Mansel. 1994. *The Early History of Radio: From Faraday to Marconi*. London, UK: Institution of Electrical Engineers.

Georgiou, M. 2006. *Diaspora, Identity and the Media: Diasporic Transnationalism and Mediated Spatialities*. New York: Hampton Press.

Ghany, Hamid. 1996. *Kamal: A Lifetime of Politics, Religion and Culture*. San Juan: Kamaluddin Mohammed.

Gilroy, Paul. 1993. *The Black Atlantic: Modernity and Double Consciousness*. Cambridge, MA: Harvard University Press.

Gooptar, Primnath. 2013a. "The Mike Men of Trinidad: East Indian Identity in Cultural, Social and Religious Activities 1935–1985". Paper presented at the conference Bonded Labour, Migration, Diaspora and Identity Formation in Historical and Contemporary Context, Paramaribo, Suriname, 6–10 June.

———. 2013b. "What Is in an Indian Diaspora Name? The Caribbean Context". *Man in India* 93 (1): 137–50.

Gordon, Barry. 2006. "My Adventure Launching TTT". *TTT Pioneers*. 15 January. http://tttpioneers.org/archives/13.

Green, Garth L. 2007. "'Come to Life': Authenticity, Value, and the Carnival as Cultural Commodity in Trinidad and Tobago". In *Identities: Global Studies in Culture and Power* 14 (1–2): 203–24. doi:10.1080/10702890601102670203.

Green, Garth L., and Philip W. Scher. 2007. *Trinidad Carnival: The Cultural Politics of a Transnational Festival*. Indianapolis: Indiana University Press.

Guilbault, Jocelyne. 2000. "Racial Projects and Musical Discourses in Trinidad, West Indies". In *Music and the Radical Imagination*, edited by Ronald Radano and Phillip V. Bohlman, 435–58. Chicago: University of Chicago Press.

Gupta, Akhil, and James Ferguson. 1997. "Culture, Power, Place: Ethnography at the End of an Era". In *Culture, Power, Place: Explorations in Critical Anthropology*, edited by Akhil Gupta and James Ferguson. 1–32. Durham, NC: Duke University Press.

Gupta, Charu. 2011. "Writing Sex and Sexuality: Archives of Colonial North India". *Journal of Women's History* 23 (4): 12–35. doi:10.1353/jowh.2011.0050.

Haan, Michael. 2005. "Numbers in Nirvana: How the 1872–1921 Indian Censuses Helped Operationalise 'Hinduism'". *Religion* 35 (1): 13–30. doi:10.1016/j.religion.2005.02.003.

Halavais, Alexander. 2000. "National Borders on the World Wide Web". *New Media and Society* 2 (1): 7–28.

Hall, Stuart. 1997. "Representation and the Media". *Media Education Foundation*. 12 March. http://www.mediaed.org/assets/products/409/.

———. 2010. "The Whites of Their Eyes: Racist Ideologies and the Media". In *Gender, Race, and Class in Media: A Critical Reader*, edited by G. Dines and J. McMahon-Humez, 81–84. Thousand Oaks, CA: Sage.

Halter, Marilyn. 2013. "Cape Verdeans and Cape Verdean Americans, 1870–1940". In *Immigrants in American History: Arrival, Adaptation, and Integration*, edited by Elliott Barkan, 269–78. Santa Barbara, CA: ABC-CLIO.

Handler, Richard. 1986. "Authenticity". *Anthropology Today* 2 (1): 2–4.

Haraksingh, Kusha. 1985. "Aspects of the Indian Experience in the Caribbean". In *Calcutta to Caroni: The East Indians of Trinidad*, edited by John La Guerre, 155–69. St Augustine: Extra Mural Studies Unit, University of the West Indies.

Hetsroni, Amir. 2005. "Rule Britannia! Britannia Rules the Waves: A Cross-Cultural Study of Five English-Speaking Versions of a British Quiz Show Format". *Communications* 30: 129–53.

Hiller, Harry H., and Tara M. Franz. 2004. "New Ties, Old Ties and Lost Ties: The Use of the Internet in Diaspora". *New Media and Society* 6 (6): 731–52.

Ho, Christine, and Keith Nurse. 2005. Introduction. In *Globalisation, Diaspora and Caribbean Popular Culture*, edited by Christine Ho and Keith Nurse, vii–xii. Kingston: Ian Randle.

Hong, Sungook. 1994. "Marconi and the Maxwellians: The Origins of Wireless Telegraphy Revisited". *Technology and Culture* 35 (4): 717–49. doi:10.2307/3106504.

Horst, Heather A. 2010. "Keeping the Link: ICTs and Jamaican Migration". In *Diasporas in the New Media Age: Identity, Politics and Community*, edited by A. Alonso and P.J. Oiarzabal, 136–50. Reno: University of Nevada Press.

Hutnyk, John. 2005. "Hybridity". *Ethnicity* 28 (1): 79–102. doi:10.1080/0141987042000280021.

———. 2006. "Culture". *Theory, Culture and Society* 23 (2–3): 351–75.

Jain, Ravindra K. 2004. "Indian Diaspora, Old and New: Culture, Class and Mobility". *Indian Anthropologist* 34 (1): 1–26.

Jayaram, Narayana. 2000. "The Dynamics of Language in Indian Diaspora: the Case of Bhojpuri/Hindi in Trinidad". *Sociological Bulletin* 49 (1): 41–62.

Jha, Jagdish Chandra. 1973. "Indian Heritage in Trinidad, West Indies". *Caribbean Quarterly* 19 (2): 28–50.

———. 1976. "The Hindu Sacraments (Rites de Passage) in Trinidad and Tobago". *Caribbean Quarterly* 22 (1): 40–52.

Johnson, Nadia Indra. 2009. "Modernizing Nationalism: Masculinity and the Performance of Anglophone Caribbean Identities". PhD dissertation, University of Miami.

Kapur, Jyotsna. 2009. "An 'Arranged Love' Marriage: India's Neoliberal Turn and the Bollywood Wedding Culture Industry". *Communication, Culture and Critique* 2(2): 221–33. doi:10.1111/j.1753-9137.2009.01036.x.

Kassim, Halima-Sa'adia. 2002. "Education and Socialization among the Indo-Muslims of Trinidad, 1917–1969". *Journal of Caribbean History* 36 (1): 100–126.

Kayum, Azeem. 2008. *Wrestling with the Goddess: A Personal Odyssey*. New York: iUniverse.

Khan, Aisha. 2009. "Sacred Subversions? Syncretic Creoles, the Indo-Caribbean, and 'Culture's In-between'". *Radical History Review* 1 (89): 165–84. doi:10.1215/01636545-2004-89-165.

Khan, Azeem. 1947. "Letter to Kamaluddin Mohammed", 9 September. Kamaluddin

Mohammed Papers. Correspondence, January 1946–October 1948. Vol. Box 1, Folder 5. Special Collections, West Indiana, Alma Jordan Library, University of the West Indies, St Augustine.

King, Richard. 1999. "Orientalism and the Modern Myth of 'Hinduism'". *Numen* 46 (2): 146–85.

Klass, Morton. 1961. *East Indians in Trinidad: A Study of Cultural Persistence*. New York: Columbia University Press.

Knowles, William H. 1959. *Trade Union Development and Industrial Relations in the British West Indies*. Berkeley and Los Angeles: University of California Press.

Korom, Frank J. 2003. *Hosay Trinidad: Muharram Performances in an Indo-Caribbean Diaspora*. Philadelphia: University of Pennsylvania Press.

Kraidy, Marwan. 2005. *Hybridity, or the Cultural Logic of Globalization*. Philadelphia: Temple University Press.

Kraidy, Marwan, and Joe Khalil. 2009. *Arab Television Industries*. London: Palgrave Macmillan and British Film Institute.

Kroeber, Alfred L., and Clyde Kluckhohn. 1952. *Culture*. New York: Meridian.

Lal, Brij V. 1998. "Understanding the Indian Indenture Experience". *South Asia: Journal of South Asian Studies* 21 (1): 215–37.

Landgraf, Edgar. 2005. "The Disintegration of Modern Culture: Nietzsche and the Information Age". *Comparative Literature* 57 (1): 25–44.

Lasky, Melvin. 2002. "The Banalization of the Concept of Culture". *Society* 39 (6): 73–81.

Lindholm, Charles. 2008. *Culture and Authenticity*. Malden, MA: Blackwell.

Liverpool, Hollis U. 1998. "Origins of Rituals and Customs in the Trinidad Carnival: African or European?" *Drama Review* 42 (3): 24–34.

———. 2001. *Rituals of Power and Rebellion: The Carnival Tradition in Trinidad and Tobago, 1763–1962*. Chicago: Research Associates School Times.

Look Lai, W. 1993. *Indentured Labor, Caribbean Sugar*. Baltimore: Johns Hopkins University Press.

Lopez, Barry. 1990. *The Rediscovery of North America*. Lexington: University Press of Kentucky.

Lorenzen, David N. 1999. "Who Invented Hinduism?" *Comparative Studies in Society and History* 41 (4): 630–59.

Lukose, Ritty A. 2007. "The Difference Diaspora Makes: Thinking Through the Anthropology of Immigrant Education in the United States". *Anthropology and Education Quarterly* 38 (4): 405–18. doi:10.1525/aeq.2007.38.4.405.

MacCannell, Dean. 1973. "Staged Authenticity: On Arrangements of Social Space in Tourist Settings". *American Journal of Sociology* 79 (3): 589–603. http://www.jstor.org/stable/2776259.

MacDonald, Scott B. 1986. *Trinidad and Tobago: Democracy and Development in the Caribbean*. New York: Praeger.

Madianou, Mirca, and Daniel Miller. 2013. *Migration and New Media: Transnational Families and Polymedia*. Oxford: Routledge, 2013.

Mahabir, Kumar. 1985. *The Still Cry: Personal Accounts of East Indians in Trinidad and Tobago during Indentureship (1845–1917)*. New York: Calaloux.

———. 1999. "The Impact of Hindi on Trinidad English". *Caribbean Quarterly* 45 (4): 13–34.

———. 2005. "Kidnappings in Trinidad: A Statistical Analysis". Paper presented at public lecture hosted by the Global Organization of People of Indian Origin, Chaguanas, Trinidad and Tobago, 3 April.

———. 2016 "Hindu and Indian-Formatted Music Radio Stations in Trinidad". *Indian Arrival Day*, 2. https://icctrinidad.wordpress.com/category/indian-arrival-day-magazine/.

Maharaj, D. Parsuram. 1998. "Back to Hinduism in Trinidad", *News India–Times*, 30 October, n.p.

Malik, Yogendra K. 1966. "The Democratic Labour Party of Trinidad: An Attempt at the Formation of a Mass Party in a Multi-Ethnic Society". PhD dissertation, University of Florida.

———. 1970. "Socio-Political Perceptions and Attitudes of East Indian Elites in Trinidad". *Western Political Quarterly* 23 (3): 552–63.

———. 1971. *East Indians in Trinidad: A Study in Minority Politics*. Oxford: Oxford University Press.

Manuel, Peter. 1997–98. "Music, Identity, and Images of India in the Indo-Caribbean Diaspora". *Asian Music* 29 (1): 17–35.

———. 2000a. *East Indian Music in the West Indies: Tan-Singing, Chutney, and the Making of Indo-Caribbean Culture*. Philadelphia: Temple University Press.

———. 2000b. "Ethnic Identity, National Identity, and Indo-Trinidadian Music". In *Music and the Racial Imagination*, edited by Ronaldo Radano and Philip V. Bohlman, 318–45. Chicago: University of Chicago Press.

Mason, Peter. 1998. *Bacchanal! The Carnival Culture of Trinidad*. Philadelphia: Temple University Press.

Mathews, Gordon. 2000. *Global Culture/Individual Identity*. London: Routledge.

McFarlane-Alvarez, Susan L. 2007. "Trinidad and Tobago Television Advertising as Third Space: Hybridity as Resistance in the Caribbean Mediascape". *Howard Journal of Communications* 18 (1): 39–55. doi:10.1080/10646170601147457.

McPhail, Thomas. 1981. *Electronic Colonialism: The Future of International Broadcasting and Communication*. Newbury Park, CA: Sage.

Meighoo, Kirk. 2008. "Ethnic Mobilisation vs. Ethnic Politics: Understanding Ethnicity in Trinidad and Tobago Politics". *Commonwealth and Comparative Politics* 46 (1): 101–27.

Merrill, John C. 1971. "The Role of the Mass Media in National Development: An Open Question for Speculation". *International Communication Gazette* 17 (4): 236–42. doi:10.1177 /0016549271017004.

Meschin, Patricia. 2004. "Piracy, Quotas, Cloud Caribbean Carnival". *Billboard*, 3 April, 37.

Miller, Daniel, and Don Slater. 2000. *The Internet: An Ethnographic Approach.* New York: Berg.

Mohammed, Kamaluddin. 2011. "At the Tables of Policy and Culture". *UWI Today*, November, 14–15.

Mohammed, P. 1988. "The 'Creolization' of Indian Women in Trinidad". In *Trinidad and Tobago: The Independence Experience, 1962–1987,* edited by Selwyn Ryan, 381–37. St Augustine, Trinidad and Tobago: Institute for Social and Economic Research.

Mohammed, Rafi. 2015. "Cultural Icon: Kamal Mohammed". *Rafimohammed.com.* 1 January. http://rafimohammed.com/cultural-icons/kamaluddin-mohammed/.

Mohammed, Shaheed Nick. 1998. "Migration and the Family in the Caribbean". *Caribbean Quarterly* 44 (3–4) (1998): 105–21.

———. 2011. *Communication and the Globalization of Culture: Beyond Tradition and Borders.* Lanham, MD: Lexington Books.

———. 2012a. "Arab and Western Images in Middle East Satellite Television Advertising". In *Advertising and Reality,* edited by Amir Hetsroni, 123–45. London: Continuum.

———. 2012b. "Home Virtual Home? A Case Study of Trinidad and Tobago Groups on Facebook". *NMEDIAC: The Journal of New Media and Culture* 8 (1). http://www .ibiblio.org/nmediac/summer2012/Articles/home_on_facebook.html.

Mohammed, Shaheed Nick, and M. Queen. 2011. "Watching Oprah in Kuwait: A Qualitative and Quantitative Investigation". *Electronic Journal of Communication* 21 (1–2). Online at http://www.cios.org/www/ejc/v21n12toc.htm.

Mohammed, Shaheed Nick, and Peer Svenkerud. 1998. "East-Indian Radio in Trinidad and Tobago: Ethnic Media in a Co-Ethnic Environment". Paper presented to the Third World Studies Conference, Omaha, Nebraska, 8–10 October.

Mohammed, Shaheed Nick, and Avinash Thombre. 2002. "East Indian Internet Radio in Trinidad and Tobago: A Discursive Cultural Practice". Paper presented at the Communication and Culture Conference, University of New Mexico. Albuquerque, New Mexico, May.

———. 2011. "Imagined Communities, Virtual Diasporas or Local Hangouts? A Study of Trinidad and Tobago Groups on Facebook". *Electronic Journal of Communication* 21 (3). Retrieved from http://www.cios.org/www/ejc/v21n34toc.htm.

———. 2014. "The Global and the Local in Trinidad and Tobago's Indian Music Format Radio". *Journal of Human Communication in the Caribbean* 1 (1): 1–32.

———. 2017. "An Investigation of User Comments on Facebook Pages of Trinidad and Tobago's Indian Music Format Radio Stations". *Journal of Radio and Audio Media* 24 (1): 111–29. doi:10.1080/19376529.2016.1252374.

Moore, Dennison. 1995. *Racial Ideology in Trinidad: The Black View of the East Indian*. Tunapuna, Trinidad and Tobago: Chakra Publishing House.

Morton, John. 1916. *John Morton of Trinidad: Pioneer Missionary of the Presbyterian Church in Canada to the East Indians in the British West Indies (Journals, Letters and Papers)*. Edited by Sarah E. Morton. Toronto: Westminster Group.

Munasinghe, Viranjini. 2001. *Callaloo or Tossed Salad?: East Indians and the Cultural Politics of Identity in Trinidad*. Ithaca: Cornell University Press.

Nettleford, Rex. 1998. "Implications for Caribbean Development". In *Caribbean Festival Arts*, edited by J.W. Nunley and J. Bettelheim, 183–98. London: University of Washington Press.

Newton, Darrell. 2008. "Calling the West Indies: The BBC World Service and Caribbean Voices". *Historical Journal of Film, Radio and Television* 28 (4): 489–97. doi:10.1080 /01439680802310308.

Nicholas, Sheilah E. 2009. "'I Live Hopi, I Just Don't Speak It': The Critical Intersection of Language, Culture, and Identity in the Lives of Contemporary Hopi Youth". *Journal of Language, Identity and Education* 8 (5): 321–34. doi:10.1080/15348450903305114.

Niranjana, Tejaswini. 2000. "Left to the Imagination: Indian Nationalisms and Female Sexuality in Trinidad". In *A Question of Silence: The Sexual Economies of Modern India*, edited by Mary E. John and Janaki Nair, 111–38. London: Zed Books.

———. 2006. *Mobilizing India: Women, Music, and Migration between India and Trinidad*. Durham, NC: Duke University Press.

Nurse, Keith. 1999. "Globalization and Trinidad Carnival: Diaspora, Hybridity and Identity in Global Culture". *Cultural Studies* 13 (4): 661–90. doi:10.1080/095023899335095.

O'Brien, Jim. 1993. "The End of Radio Silence on Internet: Author Pioneers Online Talk Radio with 'Geek of the Week'". *Computer Shopper*, 1 June, 96.

Palmer, Ruth J. 2003. "Telecommunication, Commercialism, and Boundary Crossing: The Impact on Youth and Families in Trinidad and Tobago". *Journal of Negro Education* 72 (4): 495–505.

Parasram, Jai. 2008. "GOPIO Honours Kamaluddin Mohammed". *Jyoti Communication*, 9 May. http://jyoticommunication.blogspot.com/2008/05/gopio-honours-kamaluddin -mohammed.html.

Park, Robert Ezra. 1922. *The Immigrant Press and Its Control*. New York: Harper and Brothers.

Patasar, Mungal. 1998. "The Development of Indian Music in Trinidad and Tobago". *Caribbean Dialogue* 3 (4): 67–71.

Patasar, Sharda. 2014. "Made in Trinidad, Stamped in India: Part 1". *Paradise Pulse* [online magazine], 1 December. http://www.paradisepulse.com/#!made-in-trinidad-stamped -in-india-part/c976.

Pieterse, Nederveen. 2001. "Hybridity, So What? The Anti-Hybridity Backlash and the Riddles of Recognition". *Theory, Culture and Society* (2–3): 219–45. doi:10.1177 /026327640101800211.

Potter, Simon J. 2008. "Who Listened When London Called? Reactions to the BBC Empire Service in Canada, Australia and New Zealand, 1932–1939". *Historical Journal of Film, Radio and Television* 28 (4): 475–87. doi:10.1080/01439680802310282.

Prentice, Rebecca. 2012. "'Kidnapping Go Build Back We Economy': Discourses of Crime at Work in Neoliberal Trinidad". *Journal of the Royal Anthropological Institute*, new ser., 18 (1): 45–64. doi:10.1111/j.1467-9655.2011.01730.x.

Pretelli, Matteo. 2013. "Italians and Italian-Americans 1870–1940". In *Immigrants in American History: Arrival, Adaptation, and Integration*, edited by Elliott Barkan, 437–48. Santa Barbara: ABC-CLIO.

Prorok, Carolyn V. 1997. "The Significance of Material Culture in Historical Geography: A Case Study of the Church as School in the Diffusion of the Presbyterian Mission to Trinidad". *Historical Reflections/Réflexions Historiques* 23 (3): 371–88.

Puri, Shalini. 2004. *The Caribbean Postcolonial: Social Equality, Post-Nationalism, and Cultural Hybridity*. London: Palgrave Macmillan.

Quraishi, Uzma. 2015. "Diffracted Diasporas: Trinidad's East Indians, Religio-Nationalism, and India's Independence". *Journal of Social History* 49 (2): 406–26.

Ramesar Mohan, Peggy. 1978. "Trinidad Bhojpuri: A Morphological Study". PhD dissertation, University of Michigan.

Ramnarine, Tina K. 1996. "'Indian' Music in the Diaspora: Case Studies of 'Chutney' in Trinidad and in London". *British Journal of Ethnomusicology* 5 (1): 133–53.

———. 2002. "Review of Manuel, P. 'East Indian Music in the West Indies: Tan-Singing, Chutney and the Making of Indo-Caribbean Culture'". *World of Music* 43 (2–3): 209–12.

———. 2011. "Music in Circulation between Diasporic Histories and Modern Media: exploring Sonic Politics in Two Bollywood Films Om Shanti Om and Dulha Mil Gaya". *South Asian Diaspora* 3 (2): 143–58.

Ramoutar, Paras. 1990. "Tiwari Wields Dharma with Truck and Troupe in Trinidad". *Hinduism Today*, 31 May, 1.

Rampersad, Joan. 2012. "First National Television Station". *Trinidad and Tobago Newsday*, 30 August, 14. http://www.newsday.co.tt/features/0,165549.html.

Rampersad, Kris. 2002. *Finding a Place: IndoTrinidadian Literature*. Kingston: Ian Randle.

Reddock, Rhoda. 2004. "Caribbean Masculinities and Femininities: The Impact of Globalization on Cultural Representations". In *Gender in the 21st Century: Caribbean Perspectives, Visions and Possibilities*, edited by Barbara Bailey and Elsa Leo-Rhynie, 179–216. Kingston: Ian Randle.

———. 2008. "Indian Women and Indentureship in Trinidad and Tobago 1845–1917: Freedom Denied". *Caribbean Quarterly* 54 (4): 41–68.

Reddy, Movindri. 2015. "Transnational Locality: Diasporas and Indentured South Asians". *Diaspora Studies* 8 (1): 1–17. doi:10.1080/09739572.2014.957977.

Richman, Paula. 2010. "'We Don't Change It, We Make It Applicable': Ramlila in Trinidad". *Drama Review* 54 (10): 77–104.

Rogers, Everett M. 1962. *Diffusion of Innovations*. Glencoe: Free Press.

Rohlehr, Gordon. 1998. "'We Getting the Kaiso that We Deserve': Calypso and the World Music Market". *Drama Review* 42 (3): 82–95.

Roopnarine, Lomarsh. 2006. *Indo-Caribbean Indenture: Resistance and Accommodation*. Kingston: University of the West Indies Press.

———. 2009. "Indian Social Identity in Guyana, Trinidad, and the North American Diaspora". *Wadabagei: A Journal of the Caribbean and Its Diaspora* 12 (3): 87–125.

Ross, Tara. 2014. "'Telling the Brown Stories': An Examination of Identity in the Ethnic Media of Multigenerational Immigrant Communities". *Journal of Ethnic and Migration Studies* 40 (8): 1314–29. doi:10.1080/1369183X.2013.831547.

Ryan, Selwyn. 1999. "East Indians, West Indians and the Quest for Caribbean Political Unity". *Social and Economic Studies* 48 (4): 151–84.

Sahoo, Sadananda, and Bikram K. Pattanaik. 2014. "Introduction: Diasporas in the New Global Age". In *Global Diasporas and Development: Socioeconomic, Cultural, and Policy Perspectives*, 1–16. New Delhi: Springer. doi:10.1007/978-81-322-1047-4.

Samaroo, Brinsley. 1987. "The Indian Connection: The Influence of Indian Thought and Ideas on East Indians in the Caribbean". In *Indians in the Caribbean*, edited by J.I. Singh, 25–50. London: Oriental University Press.

———. 1996. "Education as Socialization: Form and Content in the Syllabus of Canadian Presbyterian Schools in Trinidad from the Late 19th Century". *Caribbean Curriculum* 6 (1): 23–37.

Sampath, N.M. 1993. "An Evaluation of the 'Creolization' of Trinidad East Indian Adolescent Masculinity". In *Trinidad Ethnicity*, edited by K.A. Yelvington, 235–53. Knoxville: University of Tennessee Press.

Sanders, Ronald. 1978. *Broadcasting in Guyana*. London: Routledge and Kegan Paul.

Schiappa, E., P.B. Gregg, and D.E. Hewes. 2005. "The Parasocial Contact Hypothesis". *Communication Monographs* 72 (1): 92–115.

Schiller, Herbert I. 1976. *Communications and Cultural Domination*. Armonk, NY: M.E. Sharpe.

Schramm, Wilbur. 1964. *Mass Media and National Development: The Role of Information in the Developing Countries*. Redwood City, CA: Stanford University Press.

Scrase, Timothy J. 2002. "Television, the Middle Classes and the Transformation of Cultural Identities in West Bengal, India". *International Communication Gazette* 64 (4): 323–42.

Searle, Chris. 1991. "The Muslimeen Insurrection in Trinidad". *Race and Class* 33 (2): 29–43.

Shoup, G. Stanley. 1928. *Wireless Communication in the British Empire*. Trade information

Bulletin 151. Washington, DC: Bureau of Foreign and Domestic Commerce, Department of Commerce.

Singh, Jyotsna G. 1996. *Colonial Narratives, Cultural Dialogues: Discoveries of India in the Language of Colonialism*. New York: Routledge.

Singh, Kelvin. 1996. "Conflict and Collaboration: Tradition and Modernizing Indo-Trinidadian Elites (1917–56)". *New West Indian Guide/Nieuwe West-Indische Gids* 70 (3–4): 229–53.

Singh, Sherry-Ann. 2005. "Hinduism and the State in Trinidad". *Inter-Asia Cultural Studies* 6 (3): 353–65.

Sofo, Giuseppe. 2014. "Carnival, Memory and Identity". *Kultura (Skopje)* 4 (6): 17–24.

Solari, Luigi. 1948. "Guglielmo Marconi e la Marina Militare Italiana". *Rivista Marittima*, 1 February, 231.

Sollors, Werner. 1987. *Beyond Ethnicity: Consent and Descent in American Culture*. New York: Oxford University Press.

Sookdeo, Anil. 1988. "Indian West-Indians and Ethnic Processes in Trinidad and Tobago with Some Reference to the Fiji Indians". *Journal of Ethnic Studies* 16 (3): 27–45.

Stannard, David E. 1992. *American Holocaust: Columbus and the Conquest of the New World*. New York: Oxford University Press.

Stephen, Alan. 1933. "The Empire Marketing Board". *Queen's Quarterly* 40 (1): 253–60.

Straubhaar, J.D. 2008. "Global, Hybrid or Multiple? Cultural Identities in the Age of Satellite TV and the Internet". *NORDICOM Review* 29 (2): 11–29.

Sutton, Constance. 1987. "The Caribbeanization of New York City and the Emergence of a Transnational Socio-Cultural System". In *Caribbean Life in New York City: Socio-Cultural Dimensions*, by Constance R. Sutton and E. Chaney, 15–30. New York: Center for Migration Studies.

———. 2004. "Celebrating Ourselves: The Family Reunion Rituals of African-Caribbean Transnational Families". *Global Networks: A Journal of Transnational Affairs* 4 (3): 243–57. doi:10.1111/j.1471-0374.2004.00091.x.

Sydney, James. N.d. "Radio Broadcasting in Guyana". *Silver Torch*. http://silvertorch.com/g-radio-broadcasting.html.

Tanikella, Leela. 2003. "The Politics of Hybridity: Race, Gender, and Nationalism in Trinidad". *Cultural Dynamics* 15 (2): 153–81. doi:10.1177/0921374003015002002.

Teelucksingh, Jerome. 2011. "A Global Diaspora: The Indo-Trinidadian Diaspora in Canada, the United States, and England, 1967–2007". *Diaspora Studies* 4 (2): 139–54.

TATT (Telecommunications Authority of Trinidad and Tobago). 2016. *Annual Market Report 2015, Telecommunications and Broadcasting Sector*. San Juan, Trinidad and Tobago: TATT.

Tinker, H., and F. Birbalsingh. 1989. *Indenture and Exile*. Toronto: Ontario Institute for Studies in Indo-Caribbean Culture.

Tylor, Edward B. 1871. *Primitive Culture: Researches into the Development of Mythology, Philosophy, Religion, Language, Art and Custom*. London: John Murray.

UNESCO. 1983. *Tuvalu: Broadcasting Development and Training*. Paris: UNESCO.

UNFPA. 2012. *Demographics: Trinidad and Tobago*. UNFPA Caribbean. 1 January. http://caribbean.unfpa.org/public/cache/offonce/Home/Countries/TrinidadandTobago/TrinidadandTobagoDemographics;jsessionid=B46D9DC5D3271E9F9D5E969A817490oB.jahiao1.

Urla, Jacqueline. 1993. "Contesting Modernities: Language Standardization and the Production of an Ancient/Modern Basque Culture". *Critique of Anthropology* 13 (2): 101–18. doi:10.1177/0308275X9301300201.

Venugopal, Arun. 2001. "The Sounds of Home". *Little India*, 31 July, 20.

Verma, Neena. 2000. "Arrival, Survival, and Beyond Survival: The Indo-Trinidadian Journey to Political and Cultural Ascendancy". PhD dissertation, University of Toronto.

Vertovec, Steven. 1994. "'Official' and 'Popular' Hinduism in Diaspora: Historical and Contemporary Trends in Surinam, Trinidad and Guyana". *Contributions to Indian Sociology* 28 (1): 123–47. doi:10.1177/006996694028001005.

Wahab, Amar. 2011. "In the Name of Reason: Colonial Liberalism and the Government of West Indian Indentureship". *Journal of Historical Sociology* 24 (2): 209–34. doi:10.1111/j.1467-6443.2011.01396.x.

Wang, Zuoming, Joseph B. Walther, Suzanne Pingree, and Robert P. Hawkins. 2008. "Health Information, Credibility, Homophily, and Influence via the Internet: Web Sites versus Discussion Groups". *Health Communication* 23 (4): 358–68.

Warner, Keith Q. 1982. *Kaiso! The Trinidad Calypso: A Study of the Calypso as Oral Literature*. Washington, DC: Three Continents Press.

Wilson, Stacey Ann. 2012. *Politics of Identity in Small Plural Societies: Guyana, the Fiji Islands, and Trinidad and Tobago*. New York: Palgrave Macmillan.

Wilson-Heath, Carla. 1986. "Broadcasting in Kenya: Policy and Politics, 1928–1984". PhD dissertation, University of Illinois.

Yu, Sherry S. 2015. "The Inevitably Dialectic Nature of Ethnic Media". *Global Media Journal* (Canadian edition) 8 (2): 133–40.

Index

Aakash Vani 106.5 FM, 98, 104, 122–24
advertising, 7, 72, 95, 109, 148; audiences
  and markets, 2, 101, 129, 130; for Indian
  films, 85; hybridity in, 46; political,
  163–64; restrictions on, 109, 118, 120,
  164, 175; role in cultural penetration,
  79; support for Indian content, 66,
  99–101, 161
Africa, 20, 59, 61
Afro-Trinidadians, 29, 105

Black Power movement, 4, 139
Bollywood, 92, 130, 147; influence of
  films, 69–70, 125; music and songs, 6,
  37, 70, 102, 124, 160, 167
British Broadcasting Corporation, 54,
  59, 63, 64
broadcasting: early Caribbean, 58–66;
  early systems of, 53–58, ethnic,
  13–14, 74; Guyana, development of,
  22, 53–55, 59–62; international, 36, 55,
  61–65; religious, 24, 62, 66, 97, 106–10;
  Trinidad and Tobago, development of,
  2, 7, 22–23, 36, 53–65, 68–74, 79–80

Cable and Wireless Ltd, 63
cable television, 2, 80, 117, 150
calypso: acceptability to Indo-
  Trinidadians, 83, 135; as creole form,
  82, 157; popularity in the United States,
  138; radio airplay of, 87, 88, 131, 132, 161;
  social commentary in, 88, 123, 131, 132,
  135; syncretism in, 71, 123, 137, 165, 166
Canada, 2, 36, 139–41, 144, 147, 149, 155
Caribbean News Agency (CANA), 95
Carnival, 23, 49, 132, 137, 160, 166, 170
Christianity, 27, 29, 30, 33, 43, 108, 110–12,
  114, 131. See also evangelism
chutney music, 37, 70, 102, 157–60, 164–
  75; criticisms of, 102, 171–74; evolution
  and syncretism of, 23, 41, 45, 136–38,
  168–70; live events, 124, 153; origins, 23,
  166–68; radio airplay of, 102, 152, 162,
  164, 171
colonialism: broadcasting under, 55–60,
  76; "electronic", 78; global influences
  of, 3, 19, 47; legacies of, 31, 35, 40, 67;
  media under, 76–78; media after,
  78–81; race under, 43, 67, 161; religious
  biases of, 30–34, 50
content analysis, 7
cooking night, 70, 84, 166–68
creole: cultural forms, 82, 138, 166,
  168, 170; folklore, 20, 159, 160, 174;
  Mauritian (language), 24; population/
  society, 26, 28–30, 43, 47, 50, 112
cricket, 5, 20, 60, 109, 145
crusade. See evangelism
cultural authenticity, 48–51, 175
culture: 39–42; cultural assimilation, 135;

culture (*continued*)
cultural maintenance, 12, 70, 135, 156; cultural penetration, 79; development communications, 78, 79; globalization and, 39; religion and, 30

diaspora: definition and concept, 19; Indian, 23, 24, Indo-Trinidadian, 25, Indo-Trinidadian (foreign-based), 1, 7, 47, 129, 139; primary and secondary, 34–37, 140, 141, 146, 156, 173

Empire Broadcasting Service, 77
ethnic media, 12–14, 39, 160, 161
ethnicity, 5, 9, 10, 19, 22, 39, 40, 43, 44, 161. *See also* race
ethnopolitics, 22, 23, 105
evangelism, 30, 43, 115
Europe, 7, 20, 35, 63

Facebook, 7, 8, 147, 149, 153–55. *See also* social media
Fiji, 23, 35, 42

geopolitics, 62
globalization: colonialism and, 15, 18; culture and, 19, 39, 47, 48, 67–71, 134–38; historical, 47, 113, 160; identity and, 44, 45, 173; technological aspects, 36, 54–61, 141–45
Great Britain (United Kingdom): 19, 34, 57, 137, 140; broadcasting in, 54, 56, 76–78; cultural influence of, 36; independence from, 4 , 139; slavery and emancipation in 16
Guyana (British Guiana), 22–24, 34–36, 50, 53, 59–65, 69, 71, 73

Heritage Radio 101.7 FM, 104, 118, 131, 132
hierarchy of acceptability, 82, 83

Hindu Credit Union (HCU), 104, 133, 134
Hinduism: Divali, 33, 50, 51, 99, 125, 129, 131; forms and survival of, 30–34, 41, 50, 51; groups and organizations, 98, 110, 171–72; Ramayans, 123, 124; traditions and practices, 4, 20, 23, 25, 41, 71, 83, 85, 123, 131
Hot Like Pepper Radio U97.5 FM, 104, 133, 134
hybridity, 39, 45–52, 104, 134–38, 160, 164–66; critique of, 46–48, 136; cultural, 3, 5, 9, 37, 46, 158, 160; identity and, 19, 136, 165–66; musical, 2, 137, 157, 158, 172; radio programming and, 128–35, 174, 175

identity: global forces on, 6; Caribbean, 18, 22, claims, 6, 11, 19–23, 47, 119, 137–38; cultural, 5, 7, 9, 18, 37, 74; ethnic, 12, 14, 114; Hindu, 34, 113, 119; illusion of, 45; imagination, construction of, 6, 21, 36, 44; Indian, 90, 113; Indo-Trinidadian, 10, 24–34, 36, 91, 99–105; markers and cues, 34, 47; national and societal, 5, 9, 22, 39, 65, 77; post-colonial issues of, 78, 79, 136; processes and negotiation of, 14, 36–37, 51, 74; Trinidad and Tobago, 9
imagined communities, 14, 44, 173
immigration. *See* migration
Imperial Chain, 55, 56
indenture (indentureship), 3, 4, 11; descendants of labourers, 16, 19, 20; description of system, 15–18, 22; global forces influencing, 21; labourers and conditions, 33, 42–44
Irie FM1
Islam, 74, 98, 110; Eid-ul-Fitr, 97, 107, 108, 129; influence of, 67, 68; forms and survival, 30, 44, 51; Jamaat al-

Muslimeen, 93–94; radio content on, 107, 131, 150; traditions and practices, 51, 107, 131

Jamaica, 1, 63, 114, 155, 159, 169

KDKA Pittsburgh, 54, 61
KLOK (Desi 1170 AM), 14

language: among indentured labourers, 23–30, 67; Arabic, 25, 67, 68; Bhojpuri, 20, 24, 25, 26, 29, 67, 91, 159, 168, 175; devolution of, 26; English, 8, 20, 24–31, 67, 82, 86, 132, 135, 160, 168, 175; Farsi, 67, 68; Hindi, 6, 8, 24–31, 67–73, 82, 83, 123, 135, 146, 150, 168, 175; plantation Hindustani, 24, 25; Urdu, 25–27, 30, 67, 68, 72
listening curfew, 62

Marconi, Guglielmo, 141, 142
Massala 101.1 FM, 129, 145
Mastana Bahar, 90, 91
Matiabruz, 17
matikor, 166
media conglomerates, 121, 127
mic men, 84
migration, 7, 13, 19, 21, 32, 35, 36, 42, 139, 140
Mohammed, Kamaluddin , 66–74, 81, 101–2, 106
Morton, Revd James, 27–29

National Broadcasting Service, 80, 86, 98, 99
nationalism, 4, 5, 12, 19,137; diaspora and, 45; ethnicity and, 10, 19; Hindu, 31; identity and, 44; Indian, 5, 34, 113; media and, 78–79, 86, 90, 94; Trinidad and Tobago, 10, 34, 86, 112–14, 137

103 FM, 89, 93–107, 121–26, 129, 131, 134, 140–41, 145–48, 159–64, 171–75

People's Democratic Party (PDP), 112–13
People's National Movement (PNM), 35, 67, 69, 112–16, 163
The People's Station 90.5 FM, 125, 145
plantations, 3, 17, 24, 25, 28, 35, 43, 67
political independence, 5, 65; India, 5, 71, 113; identity and, 78–81, 136; Guyana, 22; media and, 78–81, 90; Trinidad and Tobago, 4, 5, 77, 112, 132; West Indian Federation, 65
propaganda, 59, 62, 63

race: colonial narratives of, 43, 161, 162; competing narratives of, 22; culture and, 41; mixed, 27; politics of, 9, 10; relations, 35, 43, 113, 114, 117
Radio 610 AM, 65, 80, 85–88, 98, 132
Radio Jaagriti, 104, 108–10; advertising restrictions, 109, 118; legal battle, 110–17; vandalized, 117
Radio Shakti. See Hot Like Pepper Radio
Radio Trinidad 730 AM: Indian music on, 72–74, 81, 106; launch, 58, 59, 62, 68; post-independence, 80; response to launch of 103 FM, 95–97
Rediffusion, 58, 64, 71, 89, 97
reggae, 1, 80, 130, 135, 158–60, 162, 164; influences of, 137; remixes, 6, 37, 124, 126, 127, 158, 160, 174

Sanatan Dharma Maha Sabha, 98, 110–16, 171
Sangeet 106.1 FM, 97–98, 122–24, 126–27
sexuality, sex, 165, 169
shortwave radio, 36, 54, 56, 58–60, 62, 73
slavery, 16, 18, 21, 47

soca, 130; acceptability among Indo-Trinidadians of, 82–83, 135–36, 160; radio airplay, 80, 135–36, 162, 164; remixes, 37, 124, 160; syncretism in, 23, 38, 41, 138, 157, 165–66
social media: audience engagement , 7, 52, 147–48, 151, 153; family ties, 139; foreign listener engagement, 155–56; messaging, 2, 7, 52, 148
streaming, 2, 7, 119, 123–25, 140–47, 154

Taj 92.3 FM, 104, 126, 127
Transcription, 63
Trinidad and Tobago Television (TTT), 79, 86, 89, 90, 92, 95, 126
Trinidad Express, 80, 81, 96
Trinidad Guardian, 80, 81, 86, 96, 115

United Nations 78, 79, 95
United States: cultural penetration from, 80; diaspora communities in, 2, 12–16, 35, 36, 139, 140, 149; military base at Chaguaramus, 89, 112; shortwave broadcasts from, 58; shortwave broadcasts to, 60; simulcasts with, 141; streaming media in, 144

VP3BG Georgetown, 60
VP3MR Georgetown, 60
VRY Georgetown, 60

WDVI Chaguaramus, 90
West Indies: cricket team, 5; Federation, 114; radio programming aimed at, 64; region, 17, 27, 61, 63, 110
WIN 101 FM, 129
wireless telegraphy, 55, 56, 142
World War I, 54
World War II, 44, 58, 62, 64, 78, 111

ZFY Georgetown, 62, 71, 73
ZQI Kingston, 63

www.ingramcontent.com/pod-product-compliance
Lightning Source LLC
Chambersburg PA
CBHW020238290326
41929CB00044B/294